Preface

Currently, around one to two billion users are able to connect to the Internet, most of them living in the industrialized parts of the world. However, if we want to improve the quality of life of the world population with the help of access to information and education, it is necessary that in the next decade an additional five billion people gain access to the Internet. Such a strategy is needed to fight poverty and inequality. The next five billion Internet users are mainly living in emerging economies. They cannot afford the kind of broadband data services many of us are enjoying today. In fact, most of them do not have internet access at all due to missing infrastructure. Therefore, the main challenge is to lower the economic barrier using new approaches for infrastructure deployment and service delivery, in order to provide affordable Internet access to billions of people living on the lower levels of the income pyramid. Lowering the costs basically means providing new low cost infrastructure, connectivity and terminals. Examples are cloud computing, advanced mobile backhaul solutions, or entry level PCs and novel mobile devices. However, this is not the only target which needs to be reached. Another aspect is to provide tailored services which are relevant to the people in their specific situation. The services should support them in their business, which is usually some kind of microentrepreneurship (e.g. farming or crafts). They also need to help overcome the lack of basic service infrastructure, e.g. by offering micro banking services. Other types of relevant mass market services are healthcare (eHealth), or services like Ambient Assisted Living. Information and Communication Technology (ICT) can help increase the well being of people thus contributing to the economic development. In turn, ICT becomes more affordable to low income communities.

Even when the cost-efficient infrastructure and the relevant applications are available, there is still the need to implement new business models to reach out to rural areas and to train the people in using those novel services. Local village entrepreneurs or village kiosk concepts are models to be deployed here. The conference of the Münchner Kreis had looked into and discuss these challenges. Speakers from the ICT industry, academia, non-governmental organizations and governmental development organizations, among them many representatives from emerging economies in Africa and Asia, have presented their activities and share positive as well as negative experiences. They had highlight the real demand for ICT, and what impact ICT creates for the wealth and lifestyle of the people. Necessary measures to improve the speed of deployment were discussed, and ultimately the business opportunities behind these concepts were been shown. Furthermore, what mature economies can learn from the emerging markets was also being discussed.

Arnold Picot   Josef Lorenz

Content

Arnold Picot • Josef Lorenz
Editors

ICT for the Next Five Billion People

Information and Communication for Sustainable Development

 Springer

Editors
Prof. Dr. Dr. Arnold Picot
Universität München
Fak. Betriebswirtschaft
Institut für Information,
Organisation und Management
Ludwigstr. 28
80539 München
Germany
picot@lmu.de

Josef Lorenz
Nokia Siemens Networks GmbH
WSE HoT
St.-Martin-Str. 53
80265 München
Germany
josef.lorenz@nsn.com

ISBN 978-3-642-12224-8   e-ISBN 978-3-642-12225-5
DOI 10.1007/978-3-642-12225-5
Springer Heidelberg Dordrecht London New York

Library of Congress Control Number: 2010928635

Cover design: WMXDesign GmbH, Heidelberg, Germany

Printed on acid-free paper

Springer is part of Springer Science+Business Media (www.springer.com)

1 Opening Remarks

Prof. Jörg Eberspächer,
Technische Universität München, Munich

A special welcome to those of you who are coming from outside Europe, from Africa, Asia and America making this conference a truly international event.

The big success of the communication technologies in the last two decades has led to more than three billion subscribers in mobile telephony and more than two billion users connected to the Internet or at least able to be connected to the Internet. Many, if not all of us, cannot live anymore without the modern digital technologies, privately and in business. However, some billions of people around the globe have not or not yet the chance to communicate electronically and to use ICT for web surfing, information retrieval, business transactions, mailing, education and so on. This is what we call the Digital Divide.

Since the economic and social welfare depends so decisively on the access to the world-wide information and ITC services, it is necessary that in the near future additional billions of people gain access to the Internet. What we want to show and discuss today in this conference are the challenges posed by this, but also some solutions.

The challenges are in three areas.

- Network and computing infrastructure – and this means primarily cheap connectivity.
- Services – especially services tailored to the needs of the people in their specific countries and situations.
- Appropriate business models – I think this is not the easiest job of the three.

To master the challenges all the different players have their own part. The ICT industry, the operators and service providers, governmental development organizations and, very important, non-governmental organizations. A lot of activities are already going on and the goal of this conference is to present some of these activities sharing the experiences and discussing what is needed to speed up the transformation processes. We believe that the development of the infrastructure, the deployment of services and applications will pose a win-win-situation for both the so called developed markets and the emerging markets. The technical and economic challenges will require and stimulate innovative ideas and lead to novel solutions which have to be contributed by creative people from all sides and from all

A. Picot and J. Lorenz (eds.), *ICT for the Next Five Billion People: Information and Communication for Sustainable Development*, DOI 10.1007/978-3-642-12225-5_1,
© Springer-Verlag Berlin Heidelberg 2010

countries. Progress in the digital world is global and all the six or, in future, seven billion people on the globe can and should benefit from it. This is one of the key messages of this conference.

This conference has been prepared by some people in the Münchner Kreis, by a program committee lead by Josef Lorenz from Nokia Siemens Networks and I thank him and his colleagues very much for the preparation.

I wish us many new insights and lively discussions. And thank you for coming here!

The first speaker is Professor Joachim von Braun. He has been director general of the International Food Policy Research Institute, IFPRI, since 2002. He guides and oversees the institute's efforts to provide research based sustainable solutions for ending hunger and malnutrition. Under his leadership IFPRI has continued to grow in food policy related strategy and governance, technology policy, markets and health nutrition policy and has significantly expanded its teams based in Africa, Asia and Latin America in response to research challenges and partners needs. Before coming to IFPRI von Braun was director of the Center for Development Research and professor of Economics and Technological Change at the University of Bonn in Germany. He has published a lot of papers and also books. One book is very important for our topic today and this is "Information and Communication Technologies for Development and Poverty Reduction". The potential of telecommunications is described discussed in this book. Hence, I can recommend it very much. Professor von Braun, the floor is yours!

2 ICT for the Poor at Large Scale: Innovative Connections to Markets and Services

Prof. Dr. Joachim von Braun,
International Food Policy Research Institute (IFPRI)[1]

The rapid spread of information and communication technologies (ICTs) has been accelerating economic and social change across all areas of human activity in past decades. Innovative technologies such as cellular telephones and wireless broadband are now reaching many parts of the developing world, including poor rural areas, and bringing in high hopes of positive development outcomes. These include accelerated economic growth, more jobs, reduced migration pressure from rural to urban areas, increased agricultural and industrial productivity, increased services and access to them, easier diffusion of innovations, and increased public administration efficiency. Yet, large inequalities remain in ICT access, both between and within countries, and large barriers continue to prevent positive outcomes for the poor. Realizing the large potential of ICTs for the poor entails a clear understanding of their needs and scaling up their connections to markets, services, and networks using ICTs.

ICTs serving the poor

Three fourths of the poor in developing countries live in rural areas and depend on agriculture-related activities. While the world is becoming more urban, the share of the poor living in rural areas remains much larger than the urban share (Figure 1). Even before the food and financial crisis hit in 2007-08, roughly 160 million people were living in ultra poverty, on less than 50 cents a day (Figure 2). The most severe deprivation has increasingly been concentrated in Sub-Saharan Africa, which has experienced a significant increase in the number of the ultra poor since 1990 and is currently home to three-quarters of the world's ultra poor people (Ahmed et al. 2007). The poor and the hungry tend to live in geographically adverse or remote areas, own few assets (including land), have limited access to credit markets, and are often excluded from social networks and political power.

---

[1] The research assistance by Bella Nestorova and Kajal Gulati and advice by Maximo Torero (all at IFPRI) for this paper is gratefully acknowledged.

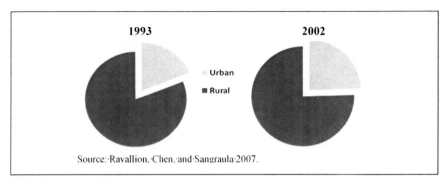

Figure 1: Urban and rural share of the poor in the developing world (%)

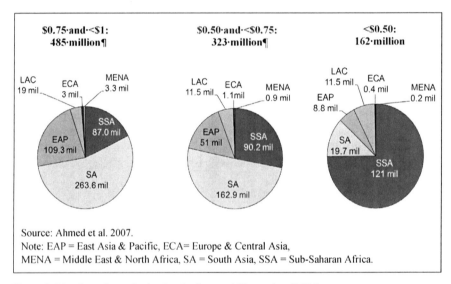

Figure 2: Number of poor in the developing world by region (2004)

Development strategies, including ICTs for development, should serve growth and the needs of the poor, as they perceive them. The development of ICTs has not been a priority in many rural areas for long because primary infrastructure and social services — such as roads, electricity, education, and health services — are in such demand. It is perhaps wrongly assumed that demand for ICTs is much lower. Actually the poor are hungry for ICT, knowing well that information serves access to education, markets, and health services. ICTs deliver clear gains for rural households. Studies of impacts of ICTs on rural households have shown a wide range of positive impacts, including:

- time and cost savings,
- better information leading to improvements in decision-making,
- greater efficiency, productivity, and diversity (Leff 1984; Tschang et al. 2002; Andrew et al. 2003),
- lower input costs, higher output prices, and information on new technologies (Gotland et al. 2004), and
- expanded market reach.

Conceptually, impacts of ICTs at the global and household level can be examined in a broad framework (Figure 3). Actual impacts of ICTs on rural households can be measured by gains in welfare, under the assumption that monetary welfare improvements eventually bring about non-monetary welfare improvements. These assessments can also be done through different methodologies: compensating variation, willingness to pay, consumption functions, and matching techniques. Using compensating variation, cases studies of Bangladesh and Peru attempt to measure the welfare effect of telephone use compared with available alternatives, such as visiting in person, sending a messenger, or sending letters (Torero and von Braun 2006). In both cases, a considerable gap was found between current prices of alternative means and local telephone use. This gap can be used as an approximation of households' willingness to pay to have telephone access. Just within the poorest quintile, the minimum estimated gains in welfare from local telephone calls compared with regular mail were US$0.11 and US$1.62 for Bangladesh and Peru, respectively. In Laos, a comparison of households that were similar in all characteristics with the exception of access to phones showed that telephone access resulted in an increase of 22 percent in per capita total consumption and 24 percent in per capita cash-based consumption. These studies illustrate that, if used properly, ICTs can benefit the poor in rural areas.

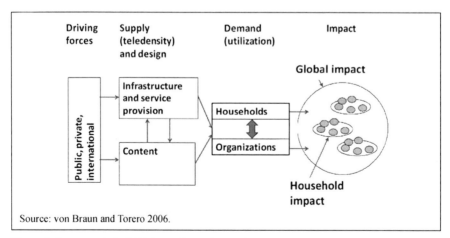

Source: von Braun and Torero 2006.

Figure 3: Framework for ICT demand, supply, and impact

Impacts on the poor: Access to markets, services, and networks

But, exactly how do ICTs have an impact on the lives and livelihoods of the poor? Research has shown that ICTs can contribute to poverty alleviation by: making markets more accessible to households; improving the quality of public goods provision, such as health services; improving the quality of human resources, primarily through education; allowing more effective utilization of existing social networks, or extending them; and creating new institutional arrangements to strengthen the rights and powers of poor people and communities. Box 1 describes several examples of how ICTs have helped the poor by providing them access to markets, services, and networks.

The reduction of the information gap at a low cost is of central importance to the poor. ICTs can be a powerful tool in removing the information asymmetries that often prevent the poor in remote areas from accessing markets, thereby leading them to lower income outcomes. ICT adoption in agriculture can allow improvements in getting timely information about prices and quality requirements, extension and latest technological know-how, and weather and water resources. Evidence from India suggests that internet kiosks that provide wholesale price information and alternative market channels to soybean farmers has led to an increase in the monthly market price by 1-5 percent. Moreover, area under soybean production has also increased significantly (Goyal 2008).

The potential role of ICTs for enhancing public health is also clear. Cross-country analysis indicates that telecommunications investment may well be associated with improved health status. Prominent applications for health include the creation of "telemedicine" centers that offer medical advice or health information to rural inhabitants. When a village chief in South Africa was asked what services he would want for his village if he had to select from a telephone line, a school, and a clinic, the answer was a "telephone line" because it would enable him to lobby ministers in the capital about the school and clinic (Micevska 2006). The demand for ICTs by the poor is not just anecdotal, but empirical evidence also indicates that a simple, linear cross-country regression of the growth rate of fixed phone lines explains about 11 percent of the growth rate variance for life expectancy. Furthermore, recent services launched in the field of mobile-banking, such as the M-PESA service introduced by Safaricom in Kenya, have brought credit services to rural areas where banking infrastructure may not be present. The service is used by 5 million customers, which is more than twice the number using traditional retail banking (BBC 2009). Services such as these provide great examples for scaling-up the role of ICTs in areas where "real" infrastructure may be missing.

Box 1: ICT enables access to markets, services, and networks

Markets:

- 1700 internet kiosks and 45 warehouses have been set up in Madhya Pradesh, India since 2000 to provide whole price information. These kiosks have offered 1-5 percent higher wholesale prices to farmers (Goyal 2008).
- Ethiopia Commodity Exchange was set up in 2008 with the help of the International Food Policy Research Institute. The institution offers new ICT-enabled market information and trading systems for connecting buyers and sellers.
- Cell phone services phased in Niger, between 2001 and 2006, provided alternative cheaper search technology to grain traders and other market actors, thereby reducing grain price dispersion by a minimum of 6.4 percent (Aker 2008).
- Cell phone adoption by fishermen in Kerala, India has provided access to different market prices and opportunities to complete market transactions without being physically present. As a result, fishermen's profits have increased by 8 percent (Jensen 2007).

Services:

- M-PESA service for transferring money using a mobile phone was launched in March 2007 in Kenya by Safaricom. It was targeted to rural people without credit cards and currently has 5 million customers (BBC 2009).
- A telemedicine system established in Alto Amazonas in Peru has seven health centers and 32 health posts. The system has led to a net savings of $320,126 over a period of four years. (Martinez et al. 2007).

Networking:

- African Virtual University is the largest network of open distance and e-learning institutions in 27 African countries. The University allows students to save money spent on tuition and learn at their own individual pace (AVU 2009).
- Indira Gandhi National Open University is reaching out to 2 million students in India and 22 other countries. It has 140 distance education institutes and is helping the disadvantaged populations (IGNOU 2009).

Entertainment and Edutainment:

- In Bangladesh, small businesses using Grameen phones have been expanded to village video and music shops, thereby becoming the entertainment hubs in the villages.
- Awareness about AIDS in India is raised using games on mobile phones, such as playing cricket with condoms. The games were downloaded by 13.5 million users, and 74 percent of the players were between ages 16-35 in 2008 (Ramey 2008).

Additionally, ICTs promote greater inclusion of individuals within networks and, even more important, increase the diversity of participants by overcoming the barriers of physical distance and social standing. They have been used for educational purposes by providing educations programs through virtual classrooms and video and audio lectures. Such services allow for significant savings to students – in some cases students save as much as 80 percent of the costs of attending in person. Two such great examples are the African Virtual University and the Indira Gandhi National Open University. The African Virtual University is the largest network of open distance and e-learning institutions in Africa, and the Indira Gandhi National Open University has 140 distance education institutes that are reaching out to two million students in India and 33 other countries (AVU 2009 and IGNOU 2009).

An often overlooked, yet crucial aspect of ICTs is the various social networking mechanisms that they offer. By providing access to a range of "fun" and "edutainment" activities such as learning about AIDS through mobile games, ICTs are playing a significant role in bridging the social rural-urban divide – an aspect that cannot be ignored in any discussions about the widening rural-urban socio-economic divide (Ramey 2008).

Some commentators hold much more skeptical views of the benefits of ICTs for development. They argue that access to ICTs largely depends on education, income, and wealth and the so-called digital divide is only a part of a much broader development divide. Limited education, inappropriate language skills, or lack of resources could prevent disadvantaged groups from accessing ICTs, ultimately exacerbating information gaps and increasing income inequality between and within countries.

Coverage and adoption: Are the poor left behind?

The variety of views about the role ICTs are also partly based on the widening digital divide between low income countries and the rest of the world. While the gap between middle income and the world seems to converging for fixed and mobile phone subscribers, the gap for the low income countries and the world is diverging from 1975 to 2005 (Figure 4). This disparity is also observed within countries. Studies based on household surveys in Peru show that the average number of telephone calls per month for the bottom income quartile is 0.5, whereas it is 6.9 calls for the top income quartile (Table 1).

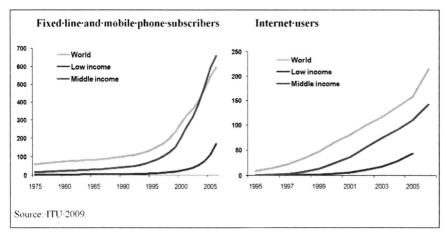

Figure 4: ICT adoption (per 1,000 people) 1995-2007

	Household income 1/	Use of public phone 2/	Avg. travel time 3/	Average call 4/	Direct monthly exp. on phone 5/
Income group					
Bottom 25%	35	65%	80	0.5	0.6
Top 25%	463	88%	27	6.9	6.2

Source: Chong, Galdo, Torero 2008.
Note: All income figures are in dollars. The exchange rate employed is 1US$=3.38S/ (World Bank, 2001)
1/ Average monthly income of the household including both farm and non-farm income in US dollars
2/ Refers to the head of the household. 3/ Walking average travel time to reach to the nearest publicly
accessible telephone in minutes. 4/ Average number of calls per month. 5/ Includes rates, only.

Table 1: Access gap within countries: Peru household survey

Even though the digital divide between countries is significant, there are still
reasons for hope. Mobile phone penetration throughout the world has been rapidly
increasing. In just the last ten years, mobile subscribers throughout the world have
been on a rise. For example, in Sub-Saharan Africa, more than 40 percent of rural
areas are now connected to mobile networks (World Bank 2008). The number of
mobile subscribers per 100 people in many developing countries has significantly
increased in just a period of seven years, from 2000 to 2007 (Figure 5). Noticeable
progress has been achieved in developing countries of Latin America, Sub-Saharan
Africa, and Asia. Potentials for providing Internet access for the "other 5 billion"
are also on the rise as 16 new satellites are planned to cover all points between 45°N
and 45°S by late 2010 (Figure 6 and Financial Times 2008).

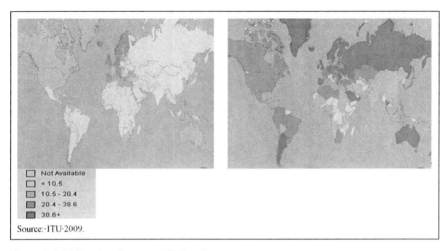

Figure 5: Mobile subscribers per 100 People

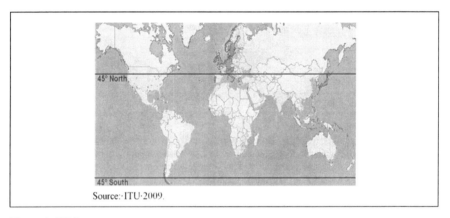

Figure 6: GSM coverage

The cost and utility of ICT expansion depend on the network size (Figure 7). Expanding the outreach of ICTs then hinges on providing an enabling environment and allowing private markets to participate. Using data from 43 countries, studies have shown that wireless phone penetration is higher in countries that have a developed telecommunication infrastructure, greater competition in markets for mobile phones, lower costs for network provision, and fewer standards for wireless penetration (Kauffman and Techatassanasoontorn 2005). However, evidence studying the expansion of mobile telephony suggests that the role of superior political institutions on diffusion is much smaller when compared to the spread of fixed line telephones and internet (Andonova and Diaz-Serrano 2008).

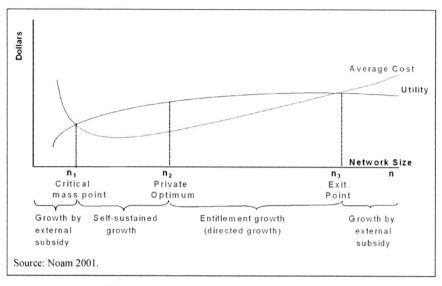

Figure 7: Costs and utility of ICT expansion

The spread of mobile phones in the absence of strong institutions becomes crucial if the growth-telecommunications links are explored. Estimates for 113 countries over a 20-year period show that a 1 percent increase in the telecommunications penetration rate might be expected to lead to a 0.03 percent increase in GDP (Torero, Chowdhury, and Bedi 2006). At the same time, models for different country groups reveal that telecommunications infrastructure has a nonlinear effect on economic output, particularly for lower and higher middle-income countries. These results imply that telecommunications networks needs to reach a critical mass to have a discernible impact on economic output. In particular, growth effects were found to be strongest in areas with telecommunications penetration rates of 5-15 percent. With the fast pace at which mobile phones are expanding in many countries throughout the world, the potential for ICTs to have growth effects looks much more optimistic.

Despite these great potentials, however, the opportunities of the digital age are not equally accessible, and the poor people have been left behind. The demand – and at time the struggle – for access by poor people is accelerating in many countries. Lack of exploitation of the opportunities that ICT holds for the developing world applies to both, the public and private sectors, as well as the community and house-hold levels.

Making ICTs pro-poor: Institutions, policies, and actions

The central question then is what *can* ICTs do and what *needs* to be done to make ICTs truly pro-poor? The lessons learnt from the many cases of ICTs for development clearly show that they are an opportunity, but not a panacea for development. For the potential benefits of ICTs to be realized in developing countries, many prerequisites need to be put in place: prompt deregulation, effective competition among service providers, free movement and adoption of technologies, targeted and competitive subsidies to reduce the access gap, and institutional arrangements to increase the use of ICTs in the provision of public goods. Successfully harnessing the power of ICT could make a substantial contribution to achieving the MDGs, both directly, through the delivery of public services, and indirectly, through the creation of new economic opportunities. Thus, looking forward, the main areas for greatest impact and expansion are:

- agricultural market information,
- micro-finance and m-banking,
- health and nutrition services,
- low-cost distance education, and
- environmental services/CO_2 markets.

Given the diverse potential benefits of ICTs, especially in the provision of public goods, subsidies traditionally used for poverty alleviation could be adapted to create incentives for the use of ICTs. For example, conditional cash transfer programs, which are largely tied to education or health, could be implemented at the community level to provide Internet access to children where educational and health services are delivered.

"3C"s are crucial for advancing ICTs for development: *connectivity, capability* to use the new tools, and relevant *content* provided in accessible and useful forms. Connectivity has been a priority, and it is a prerequisite for the other two "Cs". But given the speed at which technologies are evolving and can move –unconstrained by overly restrictive licenses and global patenting –costs could fall significantly, facilitating adoption. Hence, we should emphasize the need for all three "C's" to progress in tandem."

References

Ahmed, A., R. Hill, L. Smith, D.Wiesmann, and T. Frankenburger. 2007. *The world's most deprived: Characteristics and causes of extreme poverty and hunger.* 2020 Discussion Paper 43.Washington, D.C.: International Food Policy Research Institute.

Aker, J. 2008. Does digital divide or provide? The impact of cell phones on grain markets in Niger. Center for Development Working Paper No. 154.

Andonova, V. and L. Diaz-Serrano. 2008. Political institutions and telecommunications. *Journal of Development Economics.*

Andrew, T. N. and D. Petkov. 2003. The Need for a Systems Thinking Approach to the planning of rural telecommunications infrastructure. *Telecommunications Policy* 27: 75-93.

AVU (African Virtual University). 2009. Website. <http://www.avu.org/home.asp>. Accessed April 14, 2009.

BBC News. 2009. Help for poor to access banking. BBC < http://news.bbc.co.uk/2/hi/technology/7899396.stm>. Accessed April 14, 2009.

Chong, A., V. Galdo, and M. Torero. 2008. Access to Telephone Services and Household Income in Poor Rural Areas Using a Quasi-natural Experiment for Peru. *Economica*.

Financial Times. 2008. Google backs project to connect 3bn to net. September 9. Available at: http://www.ft.com/cms/s/0/ee2f738c-7dd0-11dd-bdbd-000077b07658.html? nclick_check=1.

Godtland, E., E. Sadoulet, A. De Janvry, R. Murgai, and O. Ortiz. 2004. The Impact of farmer field schools on Knowledge and Productivity: A study of Potato farmer in the Peruvian Andes. *Economic Development and Cultural Change* 53 (1):63-92.

Goyal, A. 2009. Information technology and rural markets: theory and evidence from a unique intervention in central India. Paper under review.

IGNOU (Indira Gandhi National Open University). 2009. Website. <http://www.ignou.ac.in>. Accessed April 14, 2009.

ITU (International Telecommunications Union). 2009. Online database. <http://www.itu.int/ITU-D/connect/gblview/> . Accessed April 14, 2009.

Jensen, R. 2007. The digital provide: information (technology), market performance, and welfare in the South Indian fisheries sector. *The Quarterly Journal of Economics* 122 (3): 879-924.

Kauffman, R., and A. Techatassanasoontorn. 2005. Is there a global digital divide for digital wireless phone technologies? *Journal of the Association for Information Systems 6: 338-382.*

Leff, N.H. 1984. Externalities, information costs, and social benefit-cost analysis for economic development: an example from telecommunications. *Economic Development and Cultural Change* 32(2): 255-276.

Martinez, A., V. Villarroel, J. Puig-Junoy, S. Joaquin, and F. del Pozo. 2007. An economic analysis of the EHAS telemedicine system in Alto Amazonas. *Journal of Telemedicine and Telecare 13(1): 7-14.*

Miceveska, M. 2005. ICT for pro-poor provision of public goods and services: a focus on health. In *Information and communication technologies for development and poverty reduction.* Ed. Torero, M. and J. von Braun. Baltimore, MD: The John Hopkins University Press for the International Food Policy Research Institute.

Noam, E. 2001. *Interconnecting the network of networks.* Cambridge, MA: MIT Press.

Ramey, C. 2008. Mobile games: learning about AIDS by playing cricket with condoms. Mobileactive.org < http://mobileactive.org/mobile-games-learning-about-aids-playing-cricket-condoms> . Accessed April 14, 2009.

Ravallion, M., S. Chen, and P. Sangraula. 2007. Evidence on the urbanization of global poverty. World Bank Policy Research Working Paper Series. Washington, DC: World Bank.

World Bank. 2008. *ICT: Connecting people and making markets work*. Washington, DC.

Torero, M. and J. von Braun. 2005. Imacts of ICt on low-income rural households. In *Information and communication technologies for development and poverty reduction*. Ed. Torero, M. and J. von Braun. Baltimore, MD: The John Hopkins University Press for the International Food Policy Research Institute.

Torero, M., S. Chowdhury, and A. Bedi 2006. Telecommunications infrastructure and economic growth: A cross-coutnry analysis. In *Information and communication technologies for development and poverty reduction*. Ed. Torero, M. and J. von Braun. Baltimore, MD: The John Hopkins University Press for the International Food Policy Research Institute.

Tschang, T., M. Chuladul, and T. Thu Le. 2002. Scaling-up information services for development. *Journal of International Development* 14: 129-41.

von Braun, J. and M. Torero. 2005. Introduction and overview. In *Information and communication technologies for development and poverty reduction*. Ed. Torero, M. and J. von Braun. Baltimore, MD: The John Hopkins University Press for the International Food Policy Research Institute.

3   People Driven Innovation –
How to create the Demand for ICT Solutions in Underserved Areas

Kazi Islam,
Grameen Solutions Limited, Dhaka, Bangladesh

Thank you for the nice introduction and I am really happy and honoured to be standing right in front of you. I come from a place, and I live among the people, and I hope to be there in the future, among the population and the challenges that all of you are here to talk about.

The question I want to pose to you is: Do you see an issue as a challenge or you can turn that around and build an opportunity out of it? Let us talk about some of the issues and then you can ask yourselves if you see them as a challenge or if you can see them as an opportunity. I will give you some examples from Bangladesh. But mind you that these are very common across most of the developing countries.

Education
In Bangladesh, for example, between 6th grades to the 10th grade student dropout rate is about 83%. There are many reasons for it. Lack of schools, the schools being expensive, parents depending on their young kids to provide support for the families and many other reasons. If you think about the female population disappearing from the education system that is also very alarming. In Bangladesh every single year there are about 100,000 female students that drop out of the education system. If you can imagine that there are about ten years worth of education, the investment, time and the effort that goes into it. You just multiply that by 100,000 that is how much investment into the education that countries like Bangladesh loose out from the disappearing female population from the education system.

Healthcare
In the healthcare, the numbers are also very common when you look across various developing countries. Infant mortality rate is about 5.4% in Bangladesh. The number of physicians for 100,000 population is 26. You can see these again as challenges. But at the same time creating more hospitals or having more doctors, are those your immediate solutions? I think you will agree that it is not. Could there be opportunities for you to leverage technology, ICT innovation to address some of this. I will show you later what we are doing in creating opportunities.

A. Picot and J. Lorenz (eds.), *ICT for the Next Five Billion People: Information and Communication for Sustainable Development*, DOI 10.1007/978-3-642-12225-5_3,
© Springer-Verlag Berlin Heidelberg 2010

Agriculture

I think we had a very interesting presentation from the previous speaker. In Bangladesh 60% of our population depend on the agriculture. Again, this is very common among developing countries. I think one of the interesting numbers that we saw is that in Bangladesh for every single dollar the consumer pays for a produce only 15% cents goes to the producer and 85% is consumed by the middleman and the inefficient systems. Could there be a system that was exampled in the previous speech where producers portion go up and the consumer's cost goes down. And, if we can do that, the people and the market will be better off. I think there are plenty of technology ideas that we can imagine.

Governance

Again, this is one of the subjects that are brought to light today. Government is so detouched from population that people – in Bangladesh if you walk around and if you ask somebody what is your proof that you are a Bangladeshi citizen? Most of them do not have any Govt. ID. If you don't have a way of communicating to your people how are you going to distribute the benefits or distribute the urgent messages? This is a tremdous challenge. I think we have seen some examples and will be getting to even more examples how we can bring government closer to the people.

Now the questions are: Do you think there is lack of innovation? Do you think there is a lack of innovative companies or lack of innovative people? Or the poor people in developing countries who dominate the global population in the world are not innovative? My answer is: No.

Let us talk about innovations. Let's think about companies like Boeing or a drug company like Eli Lilly. They are fantastic innovative companies. But think about how long it takes for a brand new Boeing to come to a country like Bangladesh. The average is about 25 years. We can argue that these are expensive planes. Countries like Bangladesh cannot afford a brand new plane. But fact remains that a new innovation from Boeing takes 25 years to come to the developing world. Similarly think about drug companies like Eli Lilly or Merk. In an average it takes about ten years for their new product to come to the developing world. Again you can argue that there are regulations and that there is a generic drug issue and all those things. But fact is that it takes ten years. A new innovation does not get to the developing world the very next day it reaches a developed country.

Those companies, the Boeing, the Eli Lilly, and other fantastic large innovative companies – if you look at their innovation process or the development process you will see a good check list. Like the product has to be at the highest quality. It has to serve in this or that geographic region. But one particular line item you may never see in an innovation process or a development process in most of the technology companies. It is the line item that says that this particular product or innovation has to be applicable and affordable in the developing countries from the day 1. So, my question is: What if we did that? What if we asked every single innovator and scientist that any time you innovate keep that component in your thought process that answers how a new innovation can be applicable and affordable in developing countries from the day 1. I think that will change a lot of things. Large companies make enough money from the developed world. What is their incentive to go beyond? Maybe we need to think about the product development process and innovation process and what we can do to bring some fundamental changes.

Are the poor not innovative? Again, as I said, my answer is no. Think about the world population. We saw the numbers. In my mind there is 80% of population that are significantly in the poor category. If you tell me that the 20% of the population who are living in the developed world are much more innovative than the 80% of the world's population, I'd argue that you are wrong.

I will give you an example of why I think that there are more innovations happening in the developing world. In Bangladesh alone every single year we have about 2.7 million people who are becoming job-ready through graduation, through becoming of age and so on. But in this single year about 700,000 people get jobs. It could be the odd job or the right job but they are employed in various sectors. That means about 2 million of the population every singe year or over 75% of the job-ready population remains unemployed. They just do not sit around and do nothing. In fact, survival instinct kicks-in and they have no other way but to be innovative to survive. In a segment of population when 75% of the population are trying to be innovative to survive I am arguing that these people have to be more innovative and they innovate more because they do survive.

A company like Microsoft or IBM are innovation machines and their innovations get to the rest of the world very fast. And, of course, they have the marketing might, and the financial might behind it. But a poor innovator in a developing world doesn't have that financial or the marketing strength. So, there has to be efforts to capture those innovations from base of the pyramid. And, in the process we will also find lots of innovations that could come from the bottom of the pyramid to the top of the pyramid. That will also eliminate various redundancies. If you go to a village in Bangladesh you will see many innovators that are creating the same thing as the villagers in Africa are creating. There is no knowledge transfer and so on as no such options exist for them.

Let us talk about the success that we can share from the Grameen side. I know a few of you know about Grameen. Dr. Yunus started Grameen almost 35 years ago and presently in Bangladesh about 26 companies carry the Grameen name. These Grameen companies vary in their service model, operating model, or management model. But, one thing these companies have in common is the focus on the poor. Every single company serves the poor.

Grameen Kalyan is a healthcare company. Currently they have healthcare clinics across Bangladesh. In Bangladesh for about two dollars a year you can buy your healthcare premium for a family with six members through Grameen Kalyan. If you think about this, it is two dollars, will this company really survive? Answer is, this company is handsomely sustainable. In Bangladesh Grameen is doing that. One catch is that you have to be really poor to get that service.

Grameen Energy is a company that provides solar home systems in Bangladesh. Again, when we think about the solar technology you don't really connect that the poor people can afford it and use it in their homes. But, in Bangladesh that is happening. Grameen Shakti, in the last seven years so far sold about 300,000 solar home systems in Bangladesh alone. The system ranges from 250 dollars to 1,000 dollars. If you think about it, 250 dollars is a tremendous amount of money for a dollar a day population. How can they really afford it? What happened is, Grameen Shakti and their people went to the villages and talked to the people. Especially, in Bangladesh we have 62% of our population that don't have any electric connections. They are not as fortunate as you are. So, they went to the villages, talked to the people and showed them that if they had the solar panels they could keep their shops open for an additional four or five hours. And hopefully that will help them have some incremental income. For example, if that additional income is $ 100 a month, will you consider setting aside about $ 10 towards the payment of your home solar systems? So, they leased those solar panels to people and provide the financing for them and as a result 300,000 people in Bangladesh have solar panels. This particular company also does biogas. In most of the rural communities in the developing countries you have lots of cattle's and the rule of thumb is if you have three cattle's in your household they will generate enough waste for you to generate enough biogas to do cooking for 3 meals a day. So, we have biogas plants popping up across Bangladesh.

Again, unless you go there, unless you talk to people how do you know what innovations are more appropriate? How do you know how your innovation really relates?

Grameen Telecom. We have heard an example of Village phone ladies. When Dr. Yunus talked about mobile phones, or the village phone ladies and phones in the hands of villagers, many really laughed at him. We thought he is crazy. People don't have enough to eat, no clothes to wear and you are talking about cell phones. But sure enough, the idea was that you will give a phone to a village lady who will rent it. It is not free. And in return she will sell the airtime to other villagers. She will earn and also pay off her rent this business really grew tremendously. In fact, that really created the revolution of mobile industry in Bangladesh. In a country with 150 million people ten years ago we had no mobile phones. Today we have more than 45 Million mobile phone users in Bangladesh.

Again, working with people can help you see better, innovate better. I think you all know about Grameen Bank, the most well-known of Grameen organisations. How did it really start? If you heard Dr. Yunus speak you might have heard the story. He believed people who are deprived of banking services need to have access to credit and to finance. So, he started studying and looking into the various models. Later on he just realised he doesn't have to work that hard. It is rather very simple. He thought what the conventional banks are doing he just needs to do the opposite. And, that is going to be his bank. His idea was that conventional banks are catering to the rich so Grameen Bank is going to cater to the poor. The conventional bank gives preference to the people with the most money. So, his bank is going to give preference to the people who have nothing. A conventional bank caters to male population. So, his bank is going to cater to female population. 97% of eight million borrowers in Grameen Bank are female. A conventional bank has lawyers and collaterals. Grameen Bank has none. Nobody as a Grameen Bank borrower had to be taken to the court. But the repayment in the Grameen Bank system remains very high which is about 99%. That is a dream for many conventional banks.

Again, being with people, coming up with ideas driven by people might be much more effective and sustainable.

Now, to my company, Grameen Solutions. Grameen has all these companies, and essentially when I came back to Bangladesh we started talking about a technology company whose focus is going to be innovating technology or identifying technology to make an impact on poverty. This is what we inspire to do. We are working with various partners in creating various solutions to make impact in poverty in rural communities. Among all devices, mobile phones happen to have largest and deepest penetration in developing countries. So, developing technology and services that leverage mobile platform is our big focus.

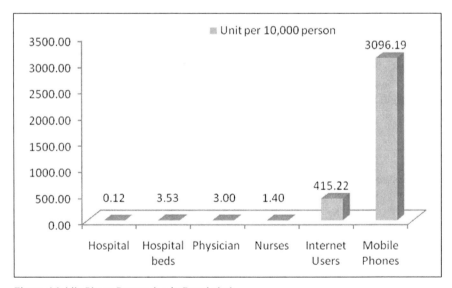

Figure: Mobile Phone Penetration in Bangladesh

This chart is very odd-looking but if you look at this comparison in Bangladesh for every 10,000 population we have about 0.12 hospitals. For every 10,000 population we have about 3.5 hospital beds. Think about the physicians, about 3for 10,000. And, you go on and the internet user is about 415.

But look at the mobile penetration for every 10,000 population. This is a very clear indication. Just think about how much access you can have if you create, if you innovate products, and the solutions, and the services based on mobile. That is where Gameen Solutions is heading to. Mobile based financial services – we have seen examples in Kenia and in Phillipines and we have studied the progress and operations very closely. Mobile Financial Services in my mind has to be the bases of mobile based trades and services. In Bangladesh for example about 88% of our financial transactions happen in cash. In US about 90% of total financial transitions are non-cash. So, if your major mode of financial transactions is cash, think about the challenges you will have. The biggest challenge you have is when you are a producer. Your biggest market opportunity is the next middleman who has cash to pay you and that is your biggest market opportunity. It doesn't matter how good the price is in another part of the country. You can't really sell it there because the mode of transaction is the major challenge. We are working with various partners to innovate financial services. Mobile based financial services will be much more acceptable and applicable in countries throughout the world. Another example, if you think about internet. We always think about how the internet has to increase in the developing countries. But think about that issues that are related to them. Number 1 is the connectivity. Connectivity in Bangladesh is still very poor. Once you have the connectivity, what you have to do is to give them a device like a PC or a desktop. A very small number of the population in developing countries has a device. Right after that, you have provide some education to teach them how to use the internet. But think about the internet and most of the contents today are in English. So, the language becomes your yet another challenge. So, there are significant challenges if you push the technology as usual.

We have plenty of technologies and ideas to create opportunities. We have seen first hand in IBM research lab, a technology called Voice Sites. Voice Sites is a concept where most of the things you do on the internet domain you are going to do that in the voice domain. May be, one day, Instead of buying a domain address you will buy a mobile phone number. Instead of writing your html code you will just speak. I am Kazi, I am an electrician and I sell light bulbs and I sell fans. As a consumer instead of writing my web address you will dial my phone number. The moment you do that the voice site says that you reached Kazi's home page and I am an electrician and these are my products and these are my services. The moment you say product it hyperlinks in the voice domain to my product page. Think about the power of that it could have in the developing countries where all these language issues and the access to content issues remain. The language barrier could be taken out right away.

Another thing is the web access through voice. While there is a huge need to be innovative in the voice domain there is already plenty of information available in the web domain. So, we are saying, can you possibly give access to people to the current content through voice. Dialling into a number and saying CNN.com will take you to the CNN.com page. By saying headline news, it just tells you what are the headlines for the day. This will most probably help lot of people in getting the benefit of the internet and the mobile based voice sites.

Mobile based healthcare is another area we are persuading very heavily. Here we are working with various technologies and organizations like mHealth Alliance. This is an alliance between UN Foundation, Rockefeller Foundation and Vodafone Foundation. In fact they are creating policies and also working with various partners. In Bangladesh we are looking into various components of mobile based health care services. The slide that you saw that there is a lack of hospitals, a lack of doctors, a lack of nurses in a country like Bangladesh and most of the developing countries. We are saying that maybe we can bring help from another dimension. Maybe we can mobilize rural health care workers and give them a technology device. And they will walk up to the villagers to everybody's doorstep and do the initial diagnosis. The final diagnosis could be just take a headache medicine and just go to sleep and you will be fine or diagnosis could be that you need to talk to a doctor and device connects you through voice or video link over your phone. We see this not as a stand alone system. We see that it has to be integrated into the health care systems in most of the developing countries. And that is something we are currently persueing. We are envisioning health city in Bangladesh where we are thinking of creating, a medical college, a medical hospital that will be connected to the regional medical hospitals. Those hospitals will be connected to maybe the village healthcare centers and finally the healthcare workers that will connect the people in their homes.

Voice translation. I think I have mentioned that the language is a huge barrier in connecting people. If you can take the language barrier out not only for in-border communication, but also for cross border communication, trades and commerce will increase tremendously. Just imagine a device that you have, a mobile phone, and you select the option that whatever communication comes your way you just want to see it or hear it in German. If it is a voice communication you want to hear it only in German. If it is a text communication you only want to read it in German. Similarly as a recipient I can also check mark in my mobile phone that whatever communication comes to me I want to hear it in Bengali. And whatever text comes in I only want to read it in Bengali. If that can be done I think it will create more opportunities in the developing world.

The financial services, health care, education and environmental issues are significant challenges. In my mind any solution that is people focussed and connecting people, helping them with their social economic development will have tremendous application in the developing world.

What can we really do?

Change
Number 1, we need to change. We talked about the thought process, the innovation process. There has to be some basic change into how you think. How you innovate. How you create. Take a different look. In Bangladesh people thought it would be land phone that would be connected trough-out the country. That didn't happen. There are about 200,000 land phones in Bangladesh and there are 45 million mobile phones. People still think in financial sector credit cards, ATM cards and internet banking will be the solution for the developing world. I don't think that is going to happen. I think it is going to be mobile based financial services because mobile has a bigger reach, a bigger penetration. So, you need to change your thinking process and your innovation process.

Get down and dirty
Next you should get down and dirty. It doesn't matter how a fantastically innovative person you are or how big innovative company you are. You are not going to make the right impact, by sitting in a big office in Berlin or Frankfurt. I ask you to come to Bangladesh, in rain and shine, get down and dirty. Know the people, know the challenges. I think your innovation will have much more relevance if you did that.

Get involved
This point is also very important. If you have a fantastic product do you think just by taking that product and standing by the road side of Dhaka you will really sell that? I don't think so. Number one is, if people cannot afford it, your product will not sell. So, you have to go there and get involved in many different ways so that socio-economic improvements happens and people's affordability increases. The second step of that is awareness. It doesn't matter how fantastic your product is. If people don't even know what that product is, how that product really makes an impact in daily personal or professional life, if they don't see and then believe the benefit of your product, how it is going to sell? So, that is the second thing you need to do. Let them know that it was innovated for them and it is made for them. Only then you can expect to see the market potential and expect your product to sell.

Build sustainability
So, this is one of two components that has to be built in every single initiative we do. Unless your initiative is sustainable on its own that initiative will collapse. Every initiative has to be sustainability built in from the day 1.

Scale
The second component is the scaling. That is also a very important component that you have to build in every single initiative for innovation in the developing world. If you can make it sustainable, if you can have a proper business model in a country

like Bangladesh, I hope that it can be taken from Bangladesh to other developing countries and you can share, your innovation to the world.

Think people first

Last point is: think people first. When we start a new company or new initiative we think about business case first. And we go trough the numbers and the financial charts. You justify the business case. But in my mind, yes, the business case is important. If you innovate the right product for the right population I think money will come. So, innovate for people first.

If you think about what really happens if you are the innovative company, or an innovative person, and if you don't think about the population in the developing world. In my mind there are two different views you can take. One from the top down. That is, you might be missing a tremendous opportunity if you are not thinking, if you are not innovating and if you are not going to the developing world. There is another look. That is the bottom up look. And that look is if you have a fantastic product, a fantastic innovation and by not taking that product to the population in the developing world you are depriving them. You are depriving them of the benefits of the fantastic products and innovations you have. In my mind from the top down you should look into how you can integrate your product and create multiple opportunities and from the bottom up I would like to see those opportunities and innovations more and more in the developing world.

APPLICATIONS FOR EMERGING ECONOMIES

Chair: Josef Lorenz, Nokia Siemens Networks, Munich

4 Transfer of services to emerging markets – mobile services, m-payment & m-health

Stanley Chia,
Vodafone Group R&D

Introduction

Transferring of services from one market to another can bring economy of scales to the development effort and help to enable service transparency for international roamers. In particular, transfer of services from developed markets to emerging markets can help to accelerate the pace of development in emerging markets and equalize the service parity between countries. Yet emerging markets have many characteristics that differentiate them from the developed world. Some are positive, such as the huge user base and strong growth potential, while others can present challenges, such as the relatively lower income level and weaker literacy rates. When considering providing or transferring services to emerging markets, the inherent market characteristics frequently demand service providers to adjust their strategic mindsets to best meet the local requirements in order to make the greatest impact.

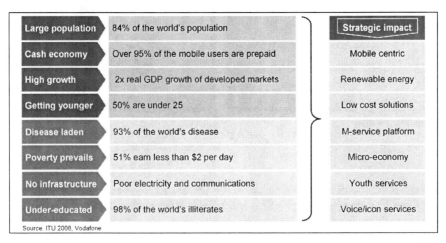

Figure 1. Market characteristics drive a different mindset for providing services to emerging markets.

A. Picot and J. Lorenz (eds.), *ICT for the Next Five Billion People: Information and Communication for Sustainable Development*, DOI 10.1007/978-3-642-12225-5_4,
© Springer-Verlag Berlin Heidelberg 2010

Within the wireless industry, the transfer of services from the developed world to emerging markets has happened for a long time and was met with different degrees of success. In this paper we picked three examples to illustrate the points. These three services are basic voice and data service, m-payment and m-health. This will help us understand the salient issues that underscore the success factors and barriers with those global undertakings.

Basic voice and data services

One of the most successful service transfers from the developed world to emerging markets in history is indeed mobile voice and text messaging services. In particular voice and text messaging have seen tremendous uptake across all emerging economies within a short period of time over the last decade. The lack of fixed infrastructure to provide adequate basic communication services has offered wireless technology-based services an unprecedented opportunity to serve the mass public, especially when the commercial and regulatory conditions are favourable. The strong uptakes of mobile subscriptions across many parts of the emerging world over the past few years are strong testimonies to this point.

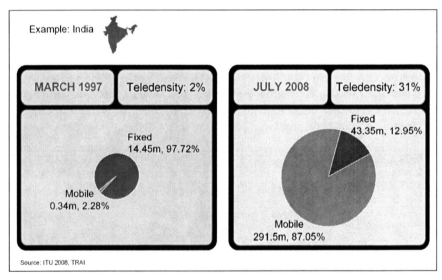

Figure 2. The successful growth of mobile communications in India.

As a whole, the phenomenal success with the transfer of basic mobile voice and data services from the developed world to emerging markets highlights that these services fulfil fundamental needs of communications by people regardless of regions. The transfer of mobile Internet based on data services in recent time is also seen to experience increasing success. Once again the inherent desire of people and businesses to access the Internet and web services has created the strong demand for mobile Internet, especially for locations where fixed infrastructure provisioning is inadequate or non-existing.

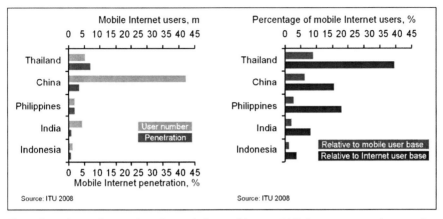

Figure 3. An increasing number of users in low and lower-middle income economies are using mobile phone as browser to access the Internet – the Mobile Internet phenomenon.

The success of the basic voice and data services indeed provided a platform for other mobile related services to evolve. However, over the history of the cellular industry, the degrees of success have not been uniform and some are more so than the others.

Mobile payment

Following the proliferation of basic mobile voice and data services into emerging markets, mobile micro-finance became the next story of success. This happens across many parts of the developing world and has benefited many countries with migrant workers working abroad as well as those with a population of low income families.

In particular, mobile payment and micro-money transfer services, as exemplified by M-PESA in Kenya, have demonstrated the popularity of micro-finance services as they tend to match the needs of the local users extremely well. This empowers people to make payment incrementally and independently in alignment with the way they are paid by their employers or customers as well as the style of spending money where many of the people are accustomed.

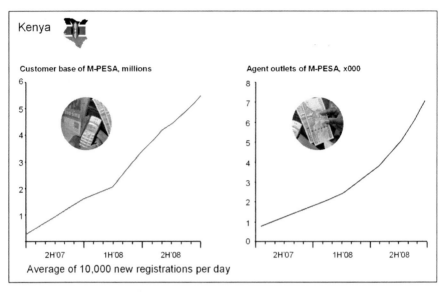

Figure 4. The strong growth of M-PESA as driven by the popularity of the service is reflected in both the growth in the customer base as well as the number of outlets where transactions can be conducted.

With the broad acceptance of mobile pre-paid services, the inherent trust to the mobile operators as honest brokers for the micro-financial transactions has greatly helped the proliferation of M-PESA. Users tend to trust the major mobile brands as a reliable and secure means to hold and transfer money to third parties. The requirement of topping-up pre-paid airtime also make people more accustom to, and develop the habit and comfort of, storing money with the mobile operator through the mobile handset in anticipation of future service utilisation.

The development of M-PESA from a concept to the first service launch took a long incubation time of multiple years and a large investment of capital in terms of millions of Euros. In addition, there were also some iterations of strategy correction of during the service development process. Although there is significant success of m-payment among the emerging markets, it is interesting to note that in the developed world, the service has not been successful at all for many of the reasons including regulatory constraints to the established habits of consumers using credit cards and the established infrastructure surrounding the payment industry.

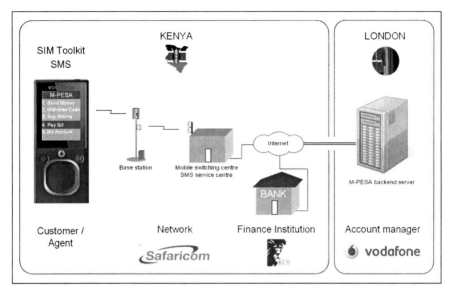

Figure 5. A simplified architecture of M-PESA showing the infrastructure within the serving country and the hosting country.

The success of a service may not always dependent on technology complexity. For instance, in the case of M-PESA, the top level architecture of the system is rather simple. The data centre is residing in the UK with the local operator in the country only needed to provide the access infrastructure and the support a network of local agents. With this architecture, the system can be easily replicated from market to market. Among others, one of the long term goals of M-PESA is, perhaps, to facilitate not only micro-money transfer within one market but to enable international micro money transfers. This could be particularly useful for countries with migrant workers and people frequently travel between them.

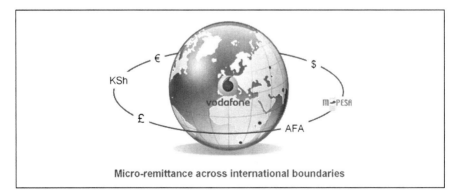

Figure 6. The concept of micro-money transfer across international boundaries.

Some of the key success factors with M-PESA included focusing on services that are useful to the mass public and that the service is reliable to gain confidence among the agents and the end-users. However, with any financial services, they are risk exposures to fraud and malpractices. These need to be closely monitored by the operator to ensure traceability of all transactions and detect any unusual pattern of activities. In addition, with any successful service, competitors will attempt to replicate the service and fight for market share sooner or later.

Mobile health

Healthcare has been a challenging topic for many emerging markets. With the proliferation of cellular infrastructure and devices, leveraging m-health to provide better healthcare to emerging markets becomes a possibility. As a whole m-health is still a challenging business across the world. Finding the right business model and monetising on the opportunity has not been easy. In the developed world, m-Health has taken a long incubation period but so far there is only very limited success. For instance, liability has been an issue that limits the proliferation of the m-health services to the mass public.

For emerging markets, m-Health may potentially play a vital role in enhancing the healthcare system, especially in transforming healthcare from a facility based service to community based system. This can enable healthcare workers to reach out to the mass public in the more remote or rural locations, where it is inconvenient for individuals to travel to clinics or hospitals that are only available in major towns and cities. Healthcare workers equipped with smart phones and intelligent software clients can bridge the gap between the scarce resource of doctors and medical experts and the vast number of patients in the field. Patient information can be sent to medical experts in the hospital for diagnosis and opinion as needed.

Together with remote diagnostics techniques and equipment, community based m-health could have a profound effect in responding quickly to potential disease outbreaks. That said, funding is still a challenge as there is generally no public healthcare network or medical insurance system among many emerging markets that can help to finance the scheme. The ability to make a profit could be even more challenging. Time will tell if sustainable business models can be found to support the community based m-health services.

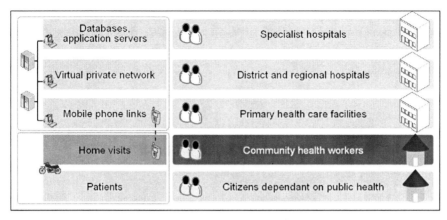

Figure 7. The concept of community based healthcare as facilitated by mobile phones

Paths of service transfer

Apart from the transfer of services from developed markets to emerging ones, the process will not be complete if service transfers among emerging markets cannot be efficiently conducted. From time to time, the regulatory regime, cultural differences and social needs could hinder the full adoption of solutions that was optimized for another market. Adjusting the service platform to meet the needs of a different country is frequently needed. The implication is that service developers have to build in flexibility in the service platform to cope with potential differences in user preference during the service development cycle.

In the long term, we would expect "reverse transfer" of services from the emerging markets back to the developed world. This may not be as easy, as by definition the developed world is supposed to be more advanced technologically and have more development resources than their emerging market counterparts in developing new services for some time to come. That said, many of the situations are common between emerging and developed markets and the former may be more motivated to optimise solutions early due to the inherent local and competitive constraints and practical needs.

Examples of situations where emerging markets could take a lead in driving tech- nology developments that are also relevant to the developed world may include the desire to reach out to areas of low population and/or the bottom of the social strata within the developed world. These have much resemblance to the conditions in many emerging markets. In addition, the need to optimise/reduce cost to provide services as well as minimise energy consumption would be generally welcomed across both emerging and developed markets alike. Though limited as it stands, it is expected more transfer of services from emerging markets to the developed world will happen in the future. This will enable both the service providers and ultimately the end-users to benefit from the investment in service developments within the emerging markets.

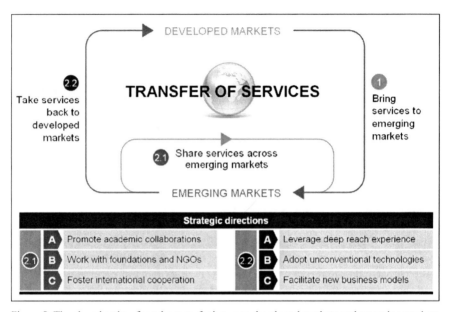

Figure 8. The closed paths of service transfer between developed markets and emerging markets and transfer between emerging markets.

To facilitate this, closer collaboration between mobile service providers with academia in both developed and emerging markets will be essential to help to better understand the needs and facilitate the development of appropriate solutions to meet local market needs. Likewise, collaboration with both government agencies and non-government organisations would also be prudent as these people have great insight into the dynamics and behaviour of users in emerging markets and can help to foster international collaborations in more meaningful ways.

Conclusion

We are seeing successful service transfers from the developed world to emerging markets where the services can fulfil basic needs of people. The general lack of fixed infrastructure in emerging markets has rendered mobile based services very popular as it can bring services to the mass public across large geographical areas quickly and, thanks to the economies of scale, they also come at relatively afford-able prices. The challenge is to enable more services to be transferred with sustain-able business models. An even bigger challenge for mobile service providers is to facilitate services to be transferred across emerging markets and even back to the developed world in the long run. If these service transfer paths can be realised, the effort of service development in both the developed and emerging worlds is expected to be even better leveraged and utilised.

References:

1. M D Bhawan, J L N Marg, "The Indian Telecom Services, Performance Indicators October-December 2008", Telecom Regulatory Authority of India.
2. V Gray, "African telecommunication / ICT indicators 2008: At a crossroads", ITU.
3. N Hughes and S Lonie, "M-PESA: Mobile Money for the "Unbanked"", February 2007, available http://www.policyinnovations.org/ideas/innovations/data/m_pesa.
4. "How we care: "Mobiling" community health", Africa medical and Research Foundation, http://www.netsquared.org/projects/how-we-care

5   Incubating Micro Enterprises in Rural South Africa –
The Use Case of Virtual Buying Cooperatives

Christian Merz,
SAP Research, Karlsruhe

I am representing SAP, the world leader in business software solutions. Typically our customers comprise companies like Coca Cola, McDonald, Nokia, Siemens, the United Nations and so on. They are not typical representatives of the economic bottom of the pyramid.

Today I want to talk about what SAP is doing to tap into the market of such bottom of the pyramid enterprises. At SAP Research we are running an umbrella research field which we call "technologies for emerging economies". Today I would like to present a concrete example of a project dealing with this research field and show-casing how solutions are differing from available products and services.

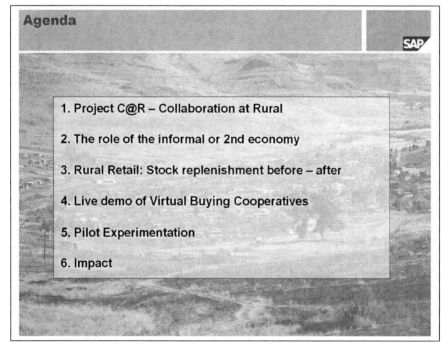

Figure 1

A. Picot and J. Lorenz (eds.), *ICT for the Next Five Billion People: Information and Communication for Sustainable Development*, DOI 10.1007/978-3-642-12225-5_5,
© Springer-Verlag Berlin Heidelberg 2010

I'm going to talk a little bit on the background of the whole project called 'Collaboration at Rural' or C@R (Fig. 1). Then it is essential to understand the role of the informal or we often call it the 2nd economy. Next we are looking at a specific use case of rural retail regarding stock replenishment. I would like to give you more insight onto the before and after scenario of stock replenishment. Then I will give a live demo of the software that we have developed. Finally I will touch our pilot experimentation 'Lessons learned' and the impact we are creating.

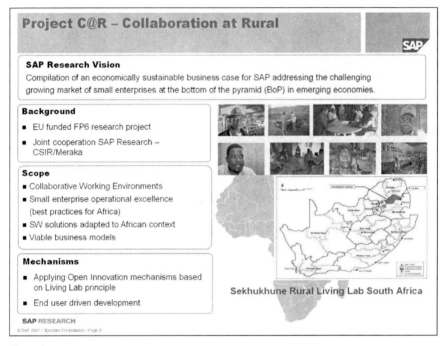

Figure 2

Coming to the background of the project (Fig. 2): our research vision is to compile an economically sustainable business case for SAP to address the bottom of the pyramid market. Of course as a profit-oriented company we're looking for solutions and products where we can make money out of. Nevertheless a sustainable business case for us could also take non-profit consideration into account as I'll explain later in more detail. The project is executed as part of the framework program 6. It is co-funded by the European Commission and we are running a joint partnership locally with the Council for Scientific and Industrial Research and the Meraka Institute in South Africa.

We are looking at software solutions that provide efficiency and effectiveness gains specifically on the small and micro enterprise level. That of course requires – and it has been often mentioned today – a local context view on the needs and requirements of the people affected. We are doing that in a living lab fashion which means that we are engaging with the end-user very closely and that we let the end user drive development. So, we carefully listen and co-design solutions together with the end user to an extent that is far beyond common practice.

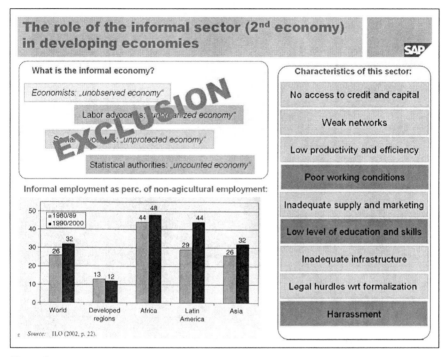

Figure 3

If you look at the world of the informal sector or the 2nd economy you get different views on how to define that sector (Fig. 3). An economist would certainly say this is an unobserved economy. A social advocate might say this is an unprotected economy. And a statistical authority would say it is an uncounted economy. In common to all definitions is exclusion. This is certainly a sector which is not properly represented in product and service delivery from the private sector. Nevertheless, if you look at the contributions in terms of employment and also in GDP the informal sector has a major impact that we have to take into account, especially in Africa, Asia and Latin America. If you look at the characteristics of such a sector you will discover that most of the informal companies, small and micro enterprises are struggling with the same kind of problems. What we do in this specific use case of virtual buying cooperatives – I am talking about today – is to tackle most of these issues like low productivity and efficiency, like inadequate supply and marketing change, like inadequate infrastructure, legal hurdles with regards to formalization and of course no access to credit and capital.

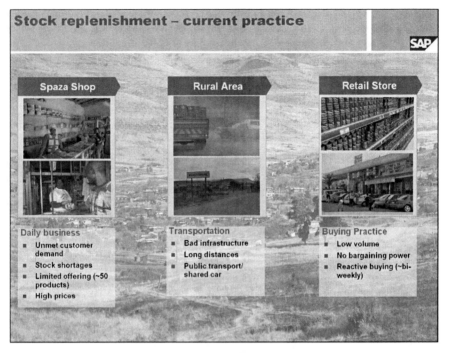

Figure 4

Looking into stock replenishment of rural retail stores so called Spaza shops (Fig. 4). We have about 100.000 Spaza Shops in South Africa. "Spaza" means hidden and stems from the former Apartheid era when these shops had to operate illegally in the townships. And still up to today they are a major backbone in terms of retail services to the local communities, in particular in rural areas. What happens is that in the rural areas in South Africa often there is a shortage of stock. The buying power on Spaza customer side is there but not met at many instances. Customers come into the shop, see the empty shells and have no chance to get the daily goods they need. This is mainly caused by infrastructure problems of rural areas as you have to overcome long distances, bad infrastructure, roads etc. So, people like the owners of these shops sometimes have to close down their shops, share a car with a neighbor or use public transport to get their stock from a retailer in the next town which is often about 60, 70 km away. You can imagine that transaction costs are very high which also results in high prices for end consumers. Dr. von Braun has already mentioned in his keynote that time is an issue to poor people. It is in fact a competitive factor that these people have to close their shop, have these high transaction costs. And they actually have no economic or bargaining power to change their situation.

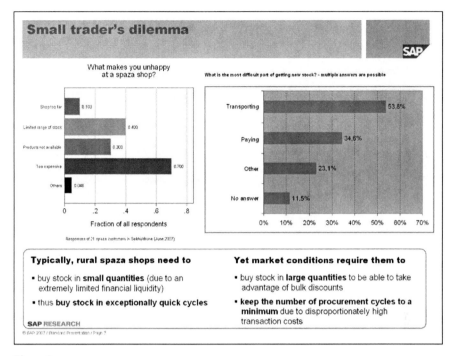

Figure 5

We have carefully looked into how we can change the situation and existing business processes. We got in contact with all people along the supply chain, listening to all stakeholders from the shop owners, the shop customers to the suppliers. We came up with identifying the key problem which we call the small trader's dilemma (Fig. 5). On the one hand side you have these rural Spaza shops that typically need to buy their stock in small quantities because they simply have a limited financial credit. Consequently they need to buy the stock in exceptional quick cycles. Often they go replenishing stock every second week. On the other side if you look on the first economy players, e.g. the big suppliers where they get their stuff from, they request market conditions to buy in large quantities and to keep the number of procurement cycles to a minimum in order to reduce transactional costs. So, there is a gap between the needs of the informal and the established economy where we as SAP for instance have a large customer base (on the established economy).

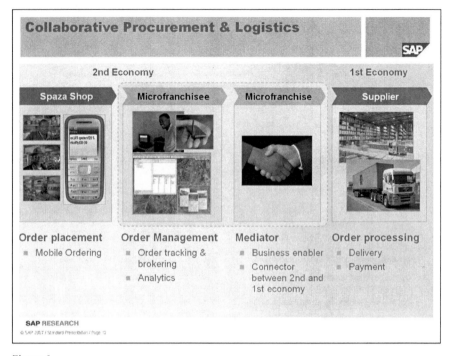

Figure 6

What have we done to change that? On the left side you have the Spaza shop owners owning a cell phone (Fig. 6). That is their only computing device. We enabled them to place an order with their mobile phone. By purpose we are using a simple, structured SMS for submitting the order – no browser based applications, no Java based applications, simply because we have to be compliant with the low end devices that are common in rural areas.

An entrepreneur acts as an intermediate service agent. We call her or him an information entrepreneur or Infopreneur. These guys are tracking the incoming orders, process them, e.g. analyze and validate and submit them as a bulk order to the suppliers. By the bulk order submission we create economies of scale that allow the 2nd economy participants to benefit from favorable business conditions granted by the 1st economy players.

Of course such business conditions need to be negotiated and we are currently doing that. Although currently not fully established we intend to run a micro franchise organization to act as a mediator between the first and second economy, e.g. taking over negotiations of business conditions. The Infopreneurs then act as micro franchisees.

Eventually the cycle is closed and the Spaza Shops get their orders delivered. This process is fully supported by the system we have developed. The system certainly does not look like the standard SAP software for those of you who have been already exposed to it. Instead we are using a geographical interface mainly because of improved usability.

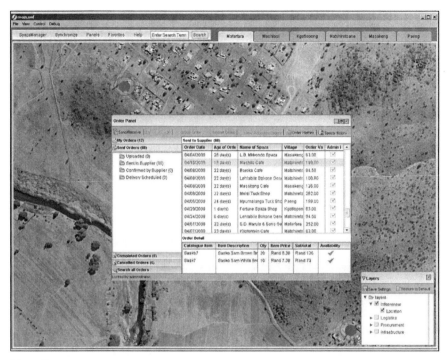

Figure 7

The process starts with the Infopreneur registering new customers, i.e. a Spaza shop (Fig. 7). The blue ones displayed are the ones who have been pre-registered but still missing some document or data to be captured. You may recognize that the Infopreneur only captures a very limited amount of data. By the way they don't store a postal address because there is no address with a street name and house number in rural South Africa. Now how do you keep track of your customers' location if there is no address? We are using geo-referenced map data to identify the GPS coordinates once the location have been picked by the Infopreneur. Then a supplier using trucks with GPS navigation systems on board knows how to find the customer. This is a very illustrative example of different requirements for software addressing the needs of emerging economies.

By the way what you see are live data. You can navigate similar to Google earth, navigate from a bird's eye viewpoint, zoom in and get the details of your customer base. What happens next? We are using a very robust solution for the shop owners for placing an order. These guys use a paper based catalogue with the products they are offered. These products have codes and of course a price and the consumption size. The structured SMS that has to be submitted for an order contains a username with a PIN and then the amount and product codes, e.g. 20 times a loaf of 700g white bread, ten times a washing powder and so on. The SMS submitted is then converted and gets into the order inbox of the Infopreneur. This is pretty much the same than checking your email inbox. The Infopreneur has different folders according to the processing stage along the supply chain. The incoming orders are listed, checked and most probably accepted according to specific criteria, e.g. taking into account granted discounts per volume, delivery duration and so on. Finally the Infopreneur synchronizes the data. This is typically an application that doesn't run according to the "always online" paradigm. It is mostly used offline and only occasionally used online. The amount of data to be synchronizes is limited to a few Kbytes per day. As the Infopreneur typically have only GPRS connectivity at its best he or she goes online once or twice a day, submit the data and the job is done. Finally after synchronization a PDF document attached to an email is sent to the supplier that gets the order listed for next day's delivery. By the way we are collaborating with the existing SAP customer Pioneer Foods in South Africa during the pilot phase.

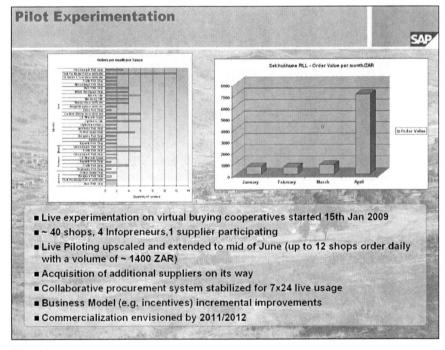

Figure 8

I am now talking about the pilot experimentation that is currently on its way (Fig. 8). We have introduced and rolled out the system in January. Since then we have gained more interest and scaled up a bit attracting more participating shops. The right chart indicates the orders' value in South African Rand as it has evolved since January. The first three months have not been too successful but now orders are gaining momentum after a few more interactive sessions with end users. I have to emphasize that the change management aspect is as important as technology solution development. You have to speak to the people. You have to provide immediate value and you have to explain the added value and train the people. Typically up to twelve shops order daily with a volume of about 120 Euro. Currently we are talking to other suppliers – also from the SAP customer base – to bring them on board and to provide delivery services to these shop owners. We are also experimenting of course with the underlying business model. A sustainable business model behind is of vital importance and we are sort of simulating incentives, e.g. for the Infopreneurs during the pilot phase. Their income for providing this kind of service is threefold. They get paid by one percentage of the order value they are processing. They get paid for newly registered customers and they get paid for proposals to improve the system usage. So, it is a highly variable incentive model behind. Another typical income stream of their current service portfolio stems from videos they are editing e.g. at weddings. This is one of their major service delivery, i.e. editing these videos and burn a DVD with some music behind and so on. What we are introducing is a sort of next level ICT service. Simply by experimenting with about 40 shops we have been enabled the Infopreneurs to make an extra earning per month of about 30 to 40 Euros.

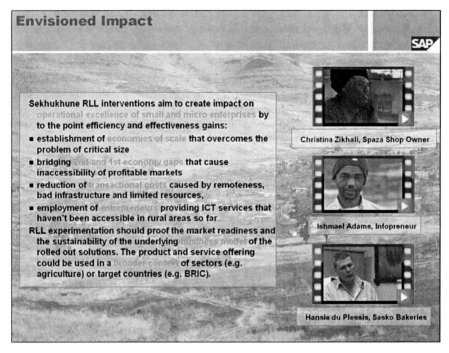

Figure 9

Now let's talk about the impact we are creating (Fig. 9). I would like to play a few videos with statements of the stakeholders that are involved, from the shop owners to the Infopreneur and to the supplier SASKO Bakery which is a branch of their holding company Pioneer Foods. We are currently working on an impact monitoring framework that we would like to apply during the experimentation period. You can imagine that reducing transactional costs improves the operational excellence of these micro-enterprises very much. Their business processes become less time consuming. Getting stock delivered significantly reduces costs compared to getting on their own and buys the stock on their own. We offer a mechanism to establish economies of scale. Micro-enterprises in rural areas simply have a problem because of scale. They have no economic power. If you bring them together and form this kind of economies of scale you achieve a sustainable bridging between the 1st economy and 2nd economy. We also experienced that the established economy has a certain interest to get more into business with the informal economy. Through the reduction of transactional costs one can even afford to introduce an entrepreneur into the supply chain who on his own wants to make some money.

We are running some other projects in South Africa. The idea is of course to extend our engagement to comparable countries like Mexico, Indonesia, countries in Latin America or Asia that have similar problems and to prove that such models can work and that there is a profitable market for the private industry.

Let us listen now what the different stakeholders say. *Interviews*

Figure 10

Finally I would like to invite you to watch the full 8 minute video that we have shot last November and that is available on the internet (Fig. 10). It gives you the whole story in a condensed movie. For more informations please send an email to me. I am happy to provide you with more material. I hope I could give you an example of how we believe we can make things differently and introduce sustainable change. Thanks a lot.

6 "What can we learn from the Developing World?"
Impact of Mobile Applications from developing Markets on Mature Economies

Prof. Gary Marsden,
University of Cape Town, South Africa

In this presentation we will explore designs that move from the developing world to the developed.

Figure 1: Google searches happening around the world

A lot of people they see Africa as in Figure 1; some kind of technology wasteland. This is an image taken from a foyer in a Google building; the length and the intensity of the coloured lines represent how many searches are happening at that instant in that place in the world. And most of the time Africa is dark.

A. Picot and J. Lorenz (eds.), *ICT for the Next Five Billion People: Information and Communication for Sustainable Development*, DOI 10.1007/978-3-642-12225-5_6,
© Springer-Verlag Berlin Heidelberg 2010

Despite this perception, technologically, Africa is an interesting place. As various other sections of this report present, Africa is not a technology wasteland. There are lots of interesting initiatives going on there. GSM networks, for instance, are highly prevalent, as can be seen in Figure 2. But does anything that happens there have any relevance back in Europe or America?

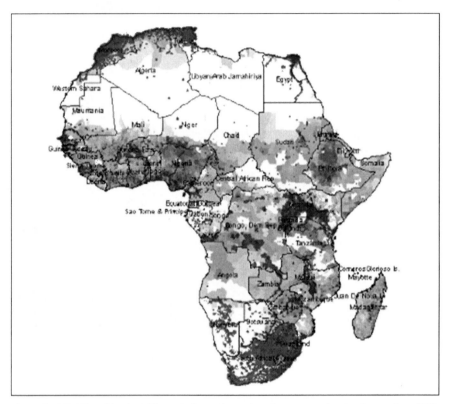

Figure 2: GSM Densities in Africa – darker colours denoting greater density. Image from ITU, 2008 (www.itu.int)

From Developing to Developed

As a technology designer originally from Europe who moved to Africa 10 years ago, I realized I had been seduced and led astray by a thing called Moore's law[1]. Essentially, Moore's law states that, for a given price, every two years technology doubles in performance. Designers in developed countries are driven to create new prototypes based on the continued improved performance of future technology. However, this is only one form of driver for design; one that is not available to designers in Africa.

Within Africa, the relative cost of technology is much higher than in developed countries, so technology adoption rates are much lower. So, rather than work on designs which require new technology to be introduced, we look at designing systems that extract more functionality from existing, or previous generation hardware.

Case Study 1 – The laptop

Looking at the example of a laptop, they have grown to become powerful and luxurious computing platforms. There are 17" laptops capable of running the latest games and even carbon fibre laptops from companies such as Ferrari and Lamborghini. Bucking this trend, however, is the XO from the One Laptop Per Child project (laptop.org) shown in Figure 3.

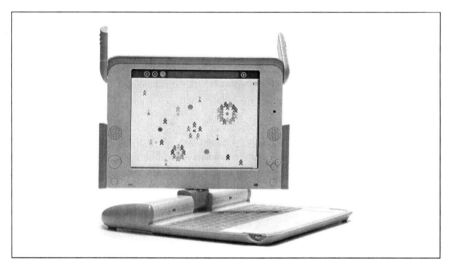

Figure 3: The XO-1 laptop from the OLPC project. Image courtesy of Mike McGregor

This is a typical device developed for emerging economies. What is interesting, however, is the effect that this device has had on laptops created for existing markets. Looking at the sales in Figure 4, it is clear that netbooks are the most popular form of laptop. Netbooks, just like the OLPC, are stripped down, low cost, robust laptops with good battery performance.

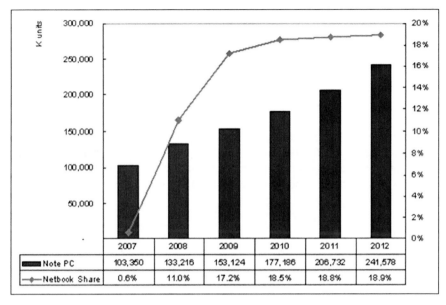

Figure 4: Existing and predicted laptop and netbook sales

Nicolas Negroponte, the driving force behind the XO project, is claiming that the design ideas for the netbooks all came from the XO. When people saw the XO they realised that it met needs that *they* had (e.g. long battery life, robust, small) much better than the high-performance laptops the manufacturers were making at that time.

Here, then, is an example of how ignoring a technology-driven design process creates a product that has relevance in other markets. By focussing on the constraints of the environment in the developing world, the designers created a product that is useful to people living in the developing world. We believe that many more worthwhile designs can be created by focussing on the technology constraints rather than assuming it is the technology that must be changed.

Case Study 2 – Photographs

The second case study resulted from our own research. We undertook an investigation which involved travelling to several countries in Africa, meeting people and seeing what they were doing with their mobile phones. One thing that became obvious is how many people had camera phones. For most people in Africa the cell phone is the only digital technology they will have access to. They won't have a digital camera. They won't have a PC. They won't have a laptop. They will just have a phone handset. The problem remains of how do they share those photographs stored on the handset. For people living in developed countries, photographs can be emailed, placed on a web site, printed or shown locally by a laptop or external monitor.

Without access to infrastructure to share photographs, many people we interviewed in Africa would have a well maintained archive on images on their handsets they used to show visitors. One person even had their wedding photographs on their handset. This led us to start thinking about what photo-management software users would want on their handsets if the handset was the only place they could store their photographs.

Firstly we wanted to create a system to share images. Given that there is no technology in the environment other than cellular handsets, we built a system that broadcasts images from one screen on to all the others within Bluetooth range [2]. See Figure 5 below.

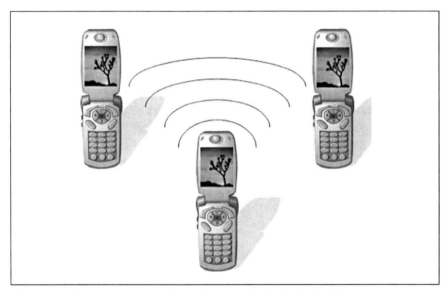

Figure 5: The image on one screen is broadcast to all others in the locality.

Again, we did not introduce new hardware, just increased the functionality of the existing technology.

We evaluated the system by sending people out for a day to take photographs with their cell phones. They then had to come back and explain, using the photographs, what had happened in their day to a friend who stayed behind. We did this with various groups of people yet at the end of the experiments the subjects did not want the experiment to end and would not give the devices back, even the ones who did have access to technologies such as the internet and laptops which the could use to share images. Again, we had created a solution that had broad appeal outside the developing world group it was intended for.

Case Study 3 – Mobile Banking

The final case study of mobile banking is widely analysed by other reports in this document. However, there is much to be learnt by looking at m-banking through the eyes of a technology designer. Mobile banking can easily fall into the trap that befalls others who wish to translate the desktop experience onto a mobile devices; they take the desktop version and try to make it fit onto a smaller screen. (Other examples would include mobile web browsing [3].) That is exactly the wrong thing to do.

Again, when we were conducting our studies in Africa we discovered purchasing behaviour that had not been reported before. For example, in a village, 200 km from the nearest tar road near the border of Zambia, Angola and DRC a trader was selling plastic basins. A customer asked for the price of the basin and then asked if he could pay in 'air time'! The trader accepted and after some pressing of buttons on cellular handsets, the customer walked off with the basin.

This is made possible as a lot of cell phone services in South Africa, or throughout Africa, have a network service that allows one user to transfer airtime credit to another network user. In effect, airtime becomes a currency. In rural communities where there is no cash this is a secure and trusted way in which to move payments around. This is backed by sale or service providers that are trusted within their community.

If we were to think about this transaction in terms of internet banking, it would take several minutes to complete – adding beneficiaries, verifying identity etc. But often, this is overkill. For example, if you were going out for a meal and someone forgot their change and asked you to lend them 5 € for a ticket home, you would be willing to hand over cash immediately. Again, a feature added to the network in Africa, originally to allow the transfer of airtime to people living rurally, has allowed the creation of a primitive bank system that does not require further technology or even legislative intervention. So successful is this facility that some 2% of the Zambian GDP is moved via airtime.

Releasing Design Ability

Finally, there are many people living in the developing world who could be creating designs and new forms of technology but do not have access to the education and resources allowing them to do this. One finds that many people who exist on the economic fringes are driven to be creative in ways that would not occur to those living in developed economies.

Figure 6: A bicycle belonging to an egg trader in Mwinilunga, Zambia.

In Figure 6, is a bicycle belonging to a trader in Zambia. To borrow the Unix/ Open Source metaphor, he has hacked his open source bicycle. He has added a pannier which will carry 75 kilos of eggs to run a delivery business. He has also upgraded the seat suspension with some soft wood so he can use this bicycle as a taxi. This is a very creative solution. We who design digital technology should be challenged and inspired by this – how can we make our technology adaptable to local conditions so that local users can create their own solutions.

These ideas are echoed in a blog called afrigadget, should you wish to see some more. Also, African design in general is celebrated at an event called "Design Indaba" which runs in the last week of February in Cape Town every year.

Conclusions

Designers in developed economies can be seduced by technology which locks them into a particular type of design thinking. However, those who are working in the developing world live under a different side of constraints which can lead to innovative solutions not obvious to those working in technology-rich environments. These solutions come about by working out what can be done with the technology that exists already. It is surprising what can be achieved.

One final thought is that there is a huge movement in developed countries towards sustainable and green design. Essentially, sustainable design is creating new things with a minimal impact. The goals of sustainable design are therefore almost identical to design for the developing world – doing more with less. There is a large potential in exploring where those design ideas overlap. Ultimately, there will be a great synergy between this sustainable movement and designing technologies for the developing world.

References:

[1] Moore, Gordon E. (1965). "Cramming more components onto integrated circuits" (PDF). Electronics Magazine, 38(8).

[2] Ah-Kun, L. & Marsden, G. (2007) "Co-Present Photo Sharing on Mobile Devices" Proceedings Mobile HCI 2007. Singapore. ACM Press. pp73-80.

[3] Jones, M. & Marsden, G. (2006) "Mobile Interaction Design" John Wiley & Sons, New York, 2006.

BUSINESS MODELS FOR SUSTAINABLE DEVELOPMENT

Chair: Christian Merz, SAP Research, Karlsruhe

7 Mobile Broadband Community Centers

Jean-Marc Cannet,
Alcatel-Lucent, Velizy

Broadband is essential to reap the social and economic benefits of Information and Communication Technologies (ICT), since it provides the fast, always-on connection needed to access the Internet. However, many economies in Asia, Latin-America, Eastern Europe, the Middle-East, and Africa, are still in the early stages of broadband adoption, with penetrations below one to two percent. Alcatel-Lucent has developed a "broadband for all" vision and enabling solutions to help service providers in high-growth economies transform their services, networks and businesses, and take advantage of the latest technologies and business models to profitably increase broadband footprint and achieve mass-market broadband growth.

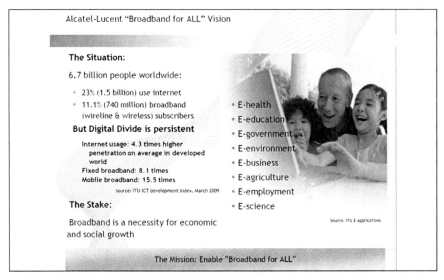

Figure 1

A. Picot and J. Lorenz (eds.), *ICT for the Next Five Billion People: Information and Communication for Sustainable Development*, DOI 10.1007/978-3-642-12225-5_7,
© Springer-Verlag Berlin Heidelberg 2010

Delivering "broadband for all" means delivering personalized and blended broadband services to advanced users, but foremost offering current Internet users affordable, mass-market broadband services for personal high-speed Internet access (Fig. 1). However, other segments are more difficult to reach: Internet users who cannot afford a personal computer and pay for a personal broadband subscription, and people who are not yet Internet literate but can benefit from Internet-based applications, such as e-government, e-business, and e-agriculture.

Figure 2

For addressing these groups that are difficult to reach, broadband community centers are very efficient delivery models (Fig. 2). These centers are also called telecenters, infocenters, information kiosks, or Internet cafes and are defined as public spots that offer communities the benefits of ICT applications through broadband access.

- Broadband community centers are cost efficient by enabling cost-sharing of PCs, modems and broadband access across users, therefore reducing the primary adoption barrier.
- Providing end-user services through broadband community centers is efficient since they enable not only self-service Internet access, but also new broadband-enabled services indirectly, through the center staff. Many offer ICT training, paving the way for future mass-market personal broadband adoption.

- Broadband community centers leverage both public and private funding efficiently by offering a mix of public services such as e-government, and private services such as desktop publishing and Internet access.

There are two types of broadband community centers, depending on the density and accessibility of the region served: in-building BCCs, generally for medium and small towns, and mobile BCCs, which travel from one village to another.

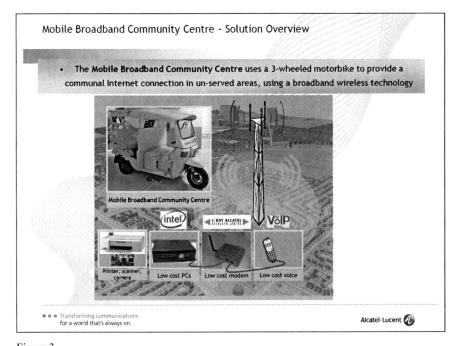

Figure 3

The mobile broadband community center uses a vehicle to provide a communal Internet connection in areas of low density or not covered efficiently by wireline broadband. It is connected through wireless broadband technology (Fig. 3).

The conditions of success towards end users

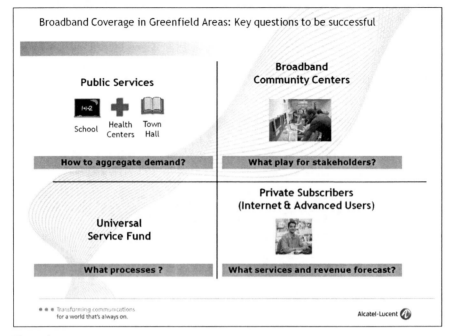

Figure 4

To ensure sufficiently high demand for services, broadband community centers need to adapt services to local needs (Fig. 4). Literacy rates, language barriers and income levels affect the services BCCs can offer in rural areas. Therefore, the following aspects should be considered:

- Locally driven content – In the local language
- Choice of self-service or staff-assisted access – Users accessing the Internet directly, or with operator assistance as needed (eSeva users interact with agents to access e-services)
- Awareness – IT is novel in rural areas. Building awareness can ensure a sustainable customer base (The Medan Infodesa Project in Malaysia adopted an integrated planning approach to create ICT awareness among the rural population)
- Affordable equipment – Given the low purchasing power of high-growth economies, the cost of personal computers is an obstacle in developing broadband community centers. Alcatel-Lucent is seeking partners, such as Intel, who are able to provide lower-cost PCs.

The incentive for service providers

Building the appropriate network architecture allows for economical broadband expansion to rural areas and enhanced service delivery in broadband community centers.

To develop the broadband footprint, no single technology fits optimally every situations. Several factors must be considered in the network design: spectrum availability, local regulations, population density, available infrastructure, geographical topology, and end-user demand. To provide affordable broadband connectivity, providers must use the right combination of broadband access technologies to address customer needs today while planning network evolution to meet the changing market demands of tomorrow. But BCCs are an opportunity for service providers to leapfrog the technology adoption cycle (latest technology is often the most cost-effective) and to enter into innovative business models.

Public-private partnership often required

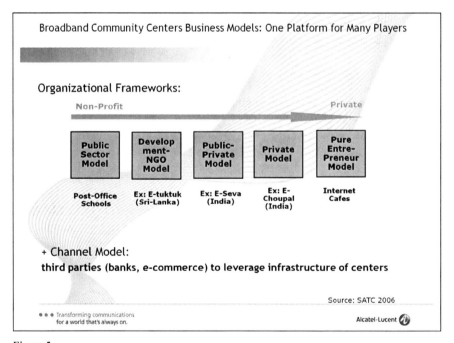

Figure 5

Broadband community centers are often based on public-private partnerships, as they benefit from public subsidies (Fig. 5). Indeed, public funding is usually essential to cover the high initial set-up costs required for a broadband community center.

The local community frequently supplies the building. And sometimes the community helps pay for the personal computer equipment. However, PCs get obsolete fast so it is important to achieve sustainable funding to ensure equipment maintenance and renewal.

Broadband community centers also have a private source of financing, as users are charged for each public service transaction and commercial service they use, such as Internet access, voice communications or desktop publishing.

The broadband community center ecosystem

The broadband community center is a complex ecosystem, involving three main stakeholders:

- Government – seeking both enhanced and more accessible public service delivery to citizens, plus operational and financial efficiencies
- Broadband community center entrepreneur – seeking a profitable business case for setting up and operating the center and providing both fee-based and free services
- Telecom service provider – seeking a profitable business case for investing in the network deployment and operation; Other stakeholders can also play a significant role
- Non-government Organizations (NGOs), and international and financial institutions such as the World Bank, the United Nations and Grameen Bank – providing financial support to bridge the digital divide and foster development in remote and underserved areas
- Local content and application providers – aimed at attracting users and making the BCC useful to various sectors of society

The case for the entrepreneur

Experience shows that the entrepreneur is key in the success of the BCC. The entrepreneur invests in and sets up the BBC, selects the implemented services to meet local end-user needs, and develops the BCC through partnerships. He or she expects financial benefits, but also expects to receive credit for helping the community.

Expected costs are staff, electricity, PCs, printers, cabling, supplies, broadband connection, depreciation and maintenance, taxes, and marketing costs to create awareness. Additional costs include the building for an in-building center, and for a mobile center, the vehicle and gasoline.

Expected revenues come from fee-based services (Internet access, voice calls, office services, business service) and handling fees or subsidies from government services.

The case for the network service provider

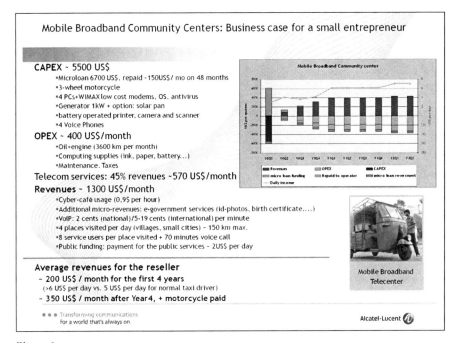

Figure 6

The expected costs for network service providers include the CAPEX and OPEX needed to expand the broadband footprint and operate the network (Fig. 6). The provider expects revenues from all additional subscribers in the area and government subsidies (from universal service fund for example) when available. Broadband community centers represent a key customer target as they allow to expand the reach of the service where a single user wouldn't have subscribed, and serves also to improve the IT-litteracy of the customers of tomorrow. Service providers can use BCC's to enter into innovative business models such as revenue sharing with the BCC entrepreneurs, through franchises or dealerships.

Conclusion

Broadband community centers are efficient delivery models to bring "broadband for all" to groups of people who cannot afford personal computers and broadband access at home, and those who are not yet Internet literate. Through broadband community centers, these populations can now benefit from ICT applications, such as e-government, e business, and e-education. Funded primarily through public and private partnerships, these broadband community centers can now efficiently expand to rural areas as new broadband technologies enable cost-effective expansion of the broadband footprint.

8   Myths about ICT for the Other Billions

Dr. Kentaro Toyama,
Microsoft Research India, Bangalore, India

Rural village with a VSAT Internet connection near Bhopal, Madhya Pradesh

Figure 1

I am going to start with a photograph of a project that was actually mentioned several times already (Fig. 1). As you can see, this is a village in the middle of India. No paved walls. There is thatched roofing. And on the roof you can see a VSAT satellite dish that is providing Internet connectivity to that village. The reason why I want to show you this photograph is just to show that especially in India there is an immense desire by agencies outside of rural areas to take technology into the villages. And, that is itself met with a lot of enthusiasm on the village side. These villagers are actually very attracted to the technology.

I think the question that many of us are asking today is represented by the two extremes expressed by these quotes. The first one says, "Kids in the developing world need the newest technology, especially really rugged hardware and innovative software." Nicholas Negroponte said this. The second one, "The world's poorest two billion people desperately need healthcare, not laptops" is from Bill Gates. These are contrasting views, but the interesting thing is that both of these people have, at least in public, moved more to the centre.

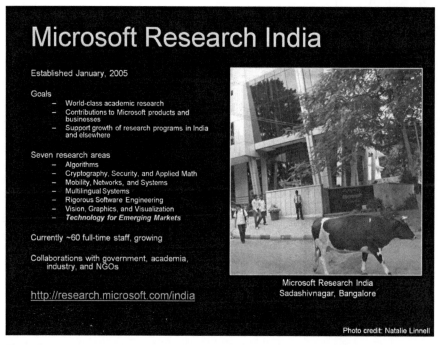

Figure 2

What I am going to do is explore what is possible in between these two quotations. Let me start first by introducing my group (Fig. 2). Here you see another photograph indicating some of the interesting conjunctions that happen in India. The lab was established in 2005 and what I am going to talk about in particular is the group that I run which is Technology for Emerging Markets. Our entire work in this group is to do research to understand how technology interacts with the developing world.

Our goals are both to understand how technology currently is either being used or not used by groups that are extremely poor and then also to find ways to invent to adopt existing technologies in such a way that they can actually serve the same communities. In our research lab we don't focus on the actual implementation of projects but we are definitely interested in having an impact, and so we try to partner with other organizations when that seems to be worthwhile. The group itself is very multidisciplinary. We have people whose backgrounds are in technology and computer science but also people who have backgrounds in anthropology, in economics, in public administration and so forth. Our methodology is quite mixed but when it goes well it goes something like what you see on the slide. So, first we spend a lot of time in immersing ourselves in a particular community, whether it is rural farmers or healthcare workers. If we spend enough time we get a certain amount of intuition about that particular community, and they will also tell us some of the problems that they are facing. If any of them appear to admit to a technology solution at a reasonable cost we will then experiment with that technology, try different designs. It alternates prototyping and testing. At the end of that cycle, if we feel that there is something that is actually having a positive impact we will move to a phase where we do a rigourous controlled trial where the goal is to measure the actual impact of a particular technology innovation. If that also goes well then our hope is to roll that out to an entity that will be interested in that particular innovation whatever it is. And they will work towards scale for larger impact. We are based in India as you can see. That small red dot represents Bangalore (Fig. 3). That is where we are based. Most of our project happen in and around Bangalore.

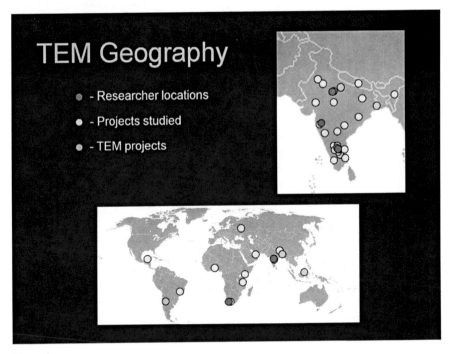

Figure 3

But we are also interested in having global impact and so we need to understand what work is done in other environents. We look at projects all around India as well as across the world.

Figure 4

These are samples of our projects (Fig. 4). You can see that we do work in education, in agriculture, in microfinance and so on. I want very briefly touch on three of our projects, more to illustrate the fact that we do quite a bit of experimentation in this area so that I then can get on to the other points that I want to get to later in this talk.

Figure 5

The first project that I am going to talk about is MultiPoint (Fig. 5). If you look at schools in developing countries, especially those run by governments quite a few are actually surprisingly beginning to have PCs and that. Their use will typically look like the one that you see in the photopraph at the top. If you count the number of children and the number of PCs you will see the ratio in this picture is six to two. Here is another photopraph where there are five children to one PC. Here is one with nine children to one PC. I have been to quite a few schools in India as well as in parts of Africa and I have not yet seen a government school where the ratio is one to one. This is a very frequent occurrence. Our technical solution to this problem is to simply provide as many mice as there are children. Sometimes, we call this project, the "one mouse per child project". You give every child a mouse. Every mouse is plugged into the same PC, via the USB interface and so the hardware is ready to go. On the software side you simply have to provide middleware that allows you to draw multiple cursers on the screen, one for each mouse. And, you can design games especially for education.

It turns out that for certain kinds of simple educational tasks like learning vocabulary or drilling mathematics, this is as effective in getting children to learn those kinds of topics compared with one PC per child. For certain kinds of educational tasks this is just as effective.

What we found is that educational ministries are usually very positive about this side because it means that if they have a fixed budget and millions of children, that this is a way that to make those numbers match.

This is a project that turns out to have very obvious commercial applications. Microsoft has taken this particular project and turned it into a software development kid that is supported and usable for free by any educational software developer. Of course, Microsoft's eventual goal is to use this as a way to provide additional incentive for schools to buy Windows and Office. Even without charging for the MultiPoint component, there is a clear benefit.

We are doing some other projects along with this. We are exploring a little bit around more around how to actually children do collaborate on particular projects as well as "Can you do this for an entire classroom?" And in fact we have been running some projects where 30 to 40 children are in a single class all with mice, all connected to a single PC. That means that you have to have a slightly different kind of game in power for the applications. But it seems to work reasonably well.

Prof. Marsden already talked about how things that you find in the developing world can have an impact on the developed world. This is one of these areas where the constraints of the developing world have forced us to find a solution that is conceivably also interesting for the developed world in the way you use PCs especially for primary schools.

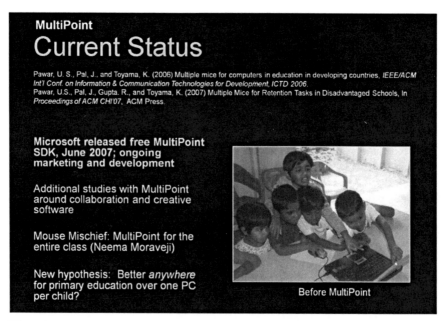

Figure 6

I just want to show you these two photographs. This photograph on the bottom right now is a situation where you see children before MultiPoint – there are little climbing all over each other to get access to the computer (Fig. 6) – and this is after MultiPoint. You see there is a dramatic difference in the way that the children are looking. And this is something we can quantify – the students are more engaged.

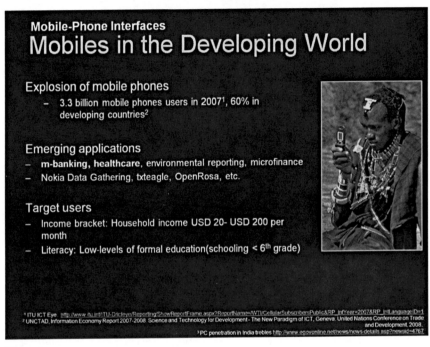

Mobile-Phone Interfaces

Mobiles in the Developing World

Explosion of mobile phones
- 3.3 billion mobile phones users in 2007[1], 60% in developing countries[2]

Emerging applications
- **m-banking, healthcare**, environmental reporting, microfinance
- Nokia Data Gathering, txteagle, OpenRosa, etc.

Target users
- Income bracket: Household income USD 20- USD 200 per month
- Literacy: Low-levels of formal education(schooling < 6th grade)

[1] ITU ICT Eye. http://www.itu.int/ITU-D/icteye/Reporting/ShowReportFrame.aspx?ReportName=/WTI/CellularSubscribersPublic&RP_intYear=2007&RP_intLanguageID=1
[2] UNCTAD, Information Economy Report 2007-2008 Science and Technology for Development- The New Paradigm of ICT, Geneva. United Nations Conference on Trade and Development, 2008.
[3] PC penetration in India trebles http://www.egovonline.net/news/news-details.asp?newsid=4767

Figure 7

The next piece of work I am going to talk about is mobile phones (Fig. 7). There has been a lot of discussion here. What we have been looking at is: How can you use a mobile phone alternatively for different kinds of tasks that you might want to use in the developing world? One of the two that I am going to talk about in particular is an interface for any kind of mobile banking application. The other is as a way to collect healthcare data.

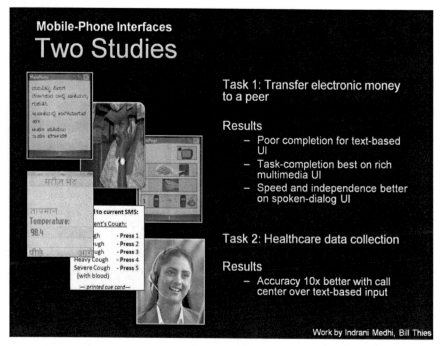

Figure 8

One of the two studies that we have conducted so far is among a text-based interface, voice-based interface where you call a voice service and then press numbers to select certain choices (Fig. 8). In a third option where there is a rich multimedia interface on the phone itself. Which one is best for use by populations that are largely illiterate? So far, what we found is that of course text-based interfaces are useless for an illiterate population. The other two have tradeoffs. On the one hand, it turns out that there is more independent task completion on a rich multimedia interface, but that the speed and acuracy are better on a spoken dialogue based interface.

In our second study, we looked at a similar problem but for a healthcare data collection. In many healthcare tasks the question is: Can you collect enough data that a government can make national level decisions about how do spend its healthcare ressources? In this case, we found that in some cases a call center that is manned by a human being is more acurate and quicker than a complex interface that somebody has to use to enter our data. One thing that we are looking at now is whether it is also cheaper as well. There are indications that in certain environments the call centers are actually a cheaper option over an automated system with a heavy server at the backend.

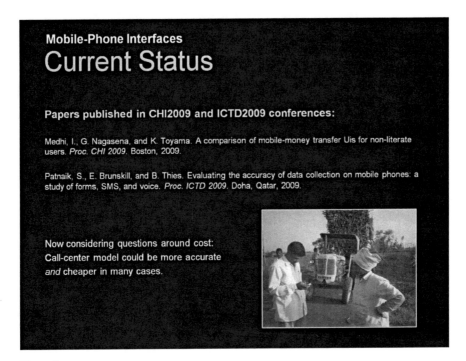

Mobile-Phone Interfaces

Current Status

Papers published in CHI2009 and ICTD2009 conferences:

Medhi, I., G. Nagasena, and K. Toyama. A comparison of mobile-money transfer Uis for non-literate users. *Proc. CHI 2009.* Boston, 2009.

Patnaik, S., E. Brunskill, and B. Thies. Evaluating the accuracy of data collection on mobile phones: a study of forms, SMS, and voice. *Proc. ICTD 2009.* Doha, Qatar, 2009.

Now considering questions around cost:
Call-center model could be more accurate
and cheaper in many cases.

Figure 9

This is a project where we believe the research paper is the major value (Fig. 9). So we published these results in peer-reviewed journals, and the hope is that over time, the community will adopt some of the learnings that we discovered.

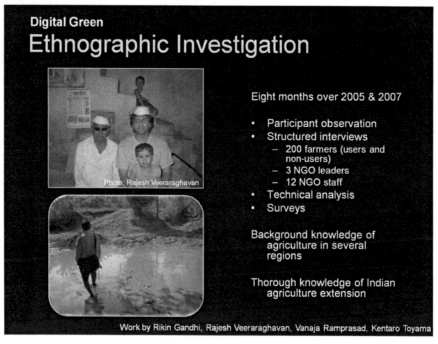

Figure 10

The third project is called Digital Green (Fig. 10). This is a project that we have run in the area of agricultural extension. Agricultural extension is the task of disseminating expert agricultural advice to the rural farmer. India itself has approximately 100.000 agricultural extension workers. They are all governmental employees. But, at that number, it's still only one agricultural extension per 2.000 farmers in the country which is just not enough to really have a significant impact. Digital Green is a system that addresses this issue. It combines two components that are absolutely critical: One is to record local farmers learning a new agricultural skill on video. On the other side, it hires local mediators whose job it is to call together sessions of farmers, watch the video with them, and periodically pause the video to ask the audience questions so that the audience remains engaged. It turns out that if you do this without the mediator no audience shows up. They don't stay even if you manage to get them there. So, it is very important to have a mediator there.

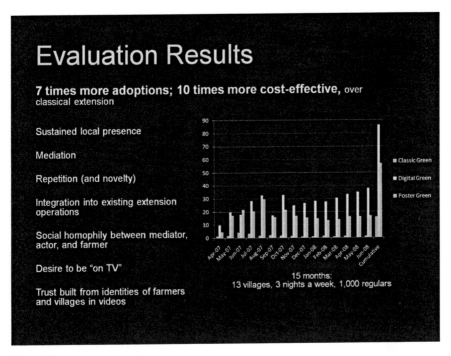

Figure 11

As it turns out this combination has an impact of helping farmers pick up new practices at a rate that is ten times more *cost effective* than the current system for agricultural extension in the country (Fig. 11). Even with the added cost of technology, the system is ten times more effective than having human beings go from door to door to try to convince farmers without video.

This is a project where we don't see any immediate revenue value to the company and so what we have decided to do is spin off a non-profit organization and have that organization receive foundation funding so that they can teach the technique to other organizations.

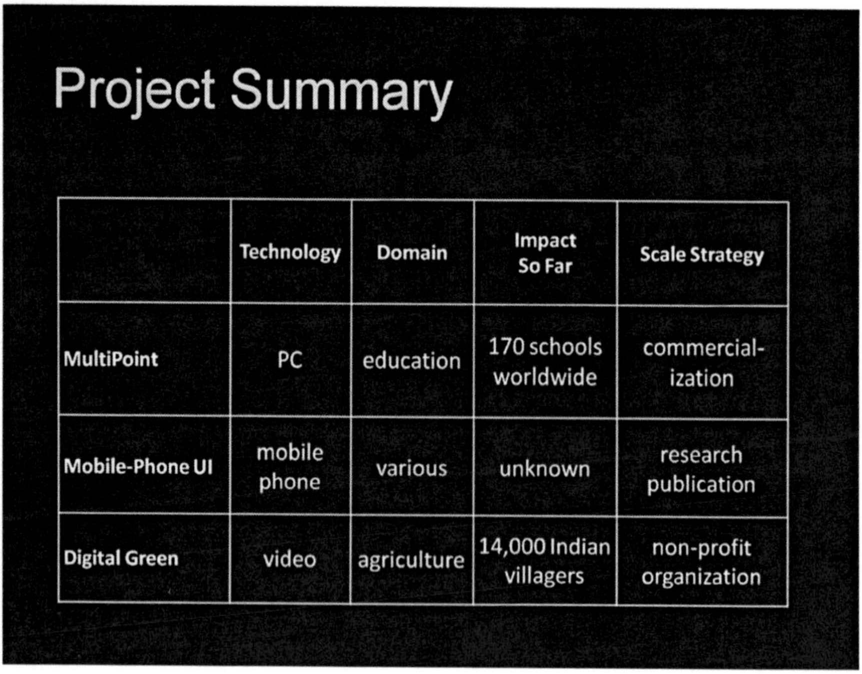

Project Summary

	Technology	Domain	Impact So Far	Scale Strategy
MultiPoint	PC	education	170 schools worldwide	commercial-ization
Mobile-Phone UI	mobile phone	various	unknown	research publication
Digital Green	video	agriculture	14,000 Indian villagers	non-profit organization

Figure 12

Just to summarize: I mentioned several different projects (Fig. 12). What I want to point out that they each use a different technology. With MultiPoint it is the PC. With the mobile-phone user interface it is a mobile phone. And, in Digital Green we are using video. They also have a very different strategy for scale. Again, with MultiPoint the goal is to have Microsoft as a commercial entity spreaded in countries that are interested. With mobile-phone research this is just a research outcome and is not in itself valuable, but for anybody who is interested in implementing a project with mobile phones the results might be helpful. Then with Digital Green our scale out strategy is to actually have a seperate non-profit organization that teaches the technique to other organizations. Again, three different business models, if you will, for each of these different projects.

One reason why I wanted to mention these projects is to show that we have done quite a bit of work with very low income populations and trying things that have a technology component. Some of them have had moderate success. Whether these will have really large success is yet to be determined.

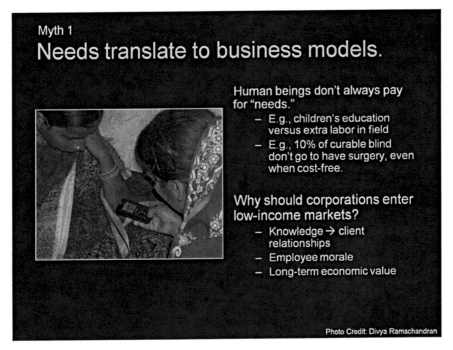

Figure 13

Based on this background I want to talk about three myths of this area that is generally called "ICT for development," or ICT4D. The first myth is that needs translate
into business models (Fig. 13). This is one of the things that I think this audience is
particular interested in hearing, which is the question of whether or not if you identify a need whether there is a real business. I think the clearest evidence that this
isn't necessarily the case is that so many of the basic needs of society are actually
not run as businesses but are effectively subsusized by the governments. This happens even in very developed countries all the way down to developing countries.
That is obviously true for education in most countries. It is largely true for healthcare, police and fire departments and so on.

Another thing that is interesting is that even in the very micro scale people don't
necessarily pay for things that you or I may proceed as needs. One of the most
interesting facts that we discovered when working with an eye care hospital in India
is that it turns out that this particular hospital operates cataract surgeries at a very
low cost. And they are able to 100% subsidize half of their patients, which means
the patients do not pay a penny for the hospital visit or for the actual cataract surgery. These are surgeries that typically cost in the developed world thousands of
dollars. In India, they can run for 20 US dollars. What's really interesting is that
when these clinics go out and find blind patients who are curable if they have the

surgery and they offer them free surgery, even then, 10% of the diagnosed patients do not make the trip to the hospital to have the surgery done. They are blind and they won't bother with the surgery so that they can see, even with all of their cost covered. And there are multiple reasons why this happens. What I really want to stress is that this is a case of something that you and I would classify as a severe need, but where the patient would not pay for it – they wouldn't even bother to do the extra trip that is required to actually have the surgery done.

There are so many such examples. They are illustrations that human beings don't always do things that are supposingly necessary for them.

This inevitably leads to motivation questions. Let us take for example Microsoft. Why do we have this lab? Why do we do our work in this area? For us it is basically three things and I think these are applicable for many large cooperations. The first is that the knowledge that you gain in doing this kind of research and potentially in implementing some of these projects it is extremely valuable in developing a relationship with clients that care about these things. For example the case of Microsoft. We are constantly asked by the governments of the world, what should they do with computers in rural areas so that the population can somehow benefit. My group gets frequently called by the other profit-oriented groups within the company to go and talk to ministers in various governmental agencies to explain to them what we have learnt about these things and then help to establish a relationship which eventually translates to a business relationship for the company. At a secondary level, there is an impact on employee morale. Again, the work that we do in our group obviously benefits to those who are doing it as this is work that we like to do. But, I have also heard many comments from other employees in the company who have nothing to do with the work that we are doing and who are still happy to see the company investing in work of this kind. This obviously has greater impact beyond the customer base. Finally, I think the long-term value of this kind of work is that especially for very large companies, the future depends on the economic growth of global markets. To the extent that we can help them grow, it ensures long-term growth for the company as well.

I would say these are at least three different reasons why a large corporation might be interested in doing work in development. It doesn't appear to have an immediate revenue benefit. On the other side, I often tell those parts of the company that are interested in working in the emerging markets that there may not be profit there. So, they need to be sure that they have another reason why they are entering those markets before they commit huge amounts of ressources to those efforts.

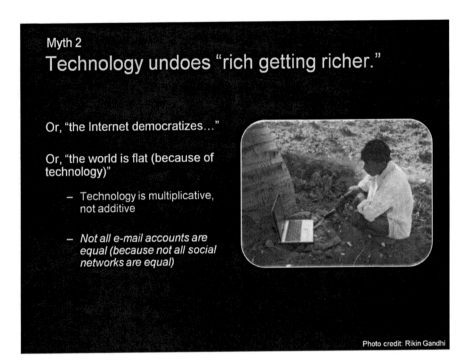

Myth 2

Technology undoes "rich getting richer."

Or, "the Internet democratizes…"

Or, "the world is flat (because of technology)"

- Technology is multiplicative, not additive

- *Not all e-mail accounts are equal (because not all social networks are equal)*

Photo credit: Rikin Gandhi

Figure 14

Myth number 2 is this idea that technology somehow counteracts the phenomenon of rich people getting richer (Fig. 14). This is a myth that is widely propagated especially because there seems to be so much proof that the Internet is a democratizing force. We all talk about how now, anybody can become a publisher of news simply be writing a blog. That must mean that Internet is a democratizing force. I think this has some truth in the developed countries. But, I think it is very untrue in developing countries. The simple example I want to give is for us to do a very simple „Gedankenexperiment" where you think about the value of having an email account. I can guarantee that everybody in this room with an email account can raise many more dollars than if I would give an email account to a very poor farmer in India. And that is not because the technology is fundamentaly different for each person. It is because all of us have different social networks. And it is the strength of that social network that is allowing the power of the email automatically shine through. One way to think about this is that the technology is actually a multiplier of what you can do without the technology. If you go to an environment in which that fundamental capability is either lacking or possibly negative, then the multiplicative factor of technology will only multiply what is not there. You won't have any positive impact.

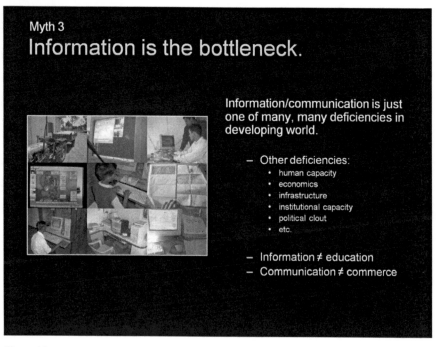

Figure 15

The third myth which we hear all the time is that somehow information is a bottleneck (Fig. 15). I have heard this statement made even by people who I think have done enough research to know that that is not quite the case. Especially in publishing environments there are so many things that are lacking that information is certainly just one of many different pieces to make something succeed. In particular, one other thing we hear all the time is that if you provide access to the Internet to everyone, they will be able to learn lots of things. They can educate themselves, and so on. Consider, all of us have access to the Internet. A good portion of MIT engineering, undergraduate curriculum is now online for all of us to access. If information equalled education, we would all be MIT engineering graduates. But that hasn't happened.

In a similar way, just providing access to the information is not sufficient for real education to happen. In many of these environments, the basic education even to absorb more information is not there. And, without that education, just providing information is effectively as empty as providing a textbook to somebody who can't read. Similarly, you can make the same case for communication but I won't go into how this actually happens in pratice.

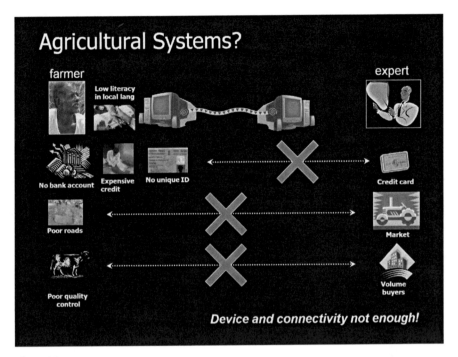

Figure 16

A lot of people talk about how we can use technology to provide better agricultural systems whether it is to connect experts with farmers (Fig. 16). The standard response is to provide Internet connection between those two people. And once you connect them magic happens: the farmer gets smarter, they automatically become richer and everybody is happy.

But, in reality what happens is that the farmer is lacking the kind of education that is really necesssary to make use of the Internet as a medium for significant exchange, whether it is literacy or even a regular former education. Similarly, if any kind of monetary exchange needs to happen, then again quite a few rural villagers are lacking the ability to both trust electronic transactions, or even to access a bank account. Of course, there are systems like M-PESA that are closing this gap. But that gap is not anyway near closed. Again, if you talk about physical transport in a lot of farming it is not enough just to have the market place information. You actually have to move goods from one place to another. If you have travelled to rural parts of developing countires, you know that transport plays a significant role. Anything that requires transport again has an additional cost. Just having the technology and the connectivity is not enough to close this gap.

You can say the same thing just about every application. E-commerce if you want to collect buyers or sellers, the same issue arise. It is not enough to have the technology and the communication. Telemedicine, another very frequent suggestion for what you can do with the Internet connection in a village, same issue all over again. These days the hype is around, a new technology, called the mobile phone … same issues.

I want to end on a note that is not entirely dark as far as technology for development. They are obviously very strong successes that are out there in the world.

I believe a lot of these myths arise because we are judging these projects from a different frame of mind than one that would be appropriate for the developing context. Here is a nice visual representation of digital divide. You need many different components to actually bridge that divide. It is not just a digital divide by itself. It requires physical, human, social, and economic components.

In the developed world, you have a good portion of this already set up. When the digital component comes, people can take advantage of it and magic does happen.

In the developing world the rest is not there to begin with. So, you can provide the additional digital component, but by itself, it doesn't bridge the gap. The message for those of us who are really interested in closing this gap is that you have to work with organizations that are fundamentally interested and capable of providing those other entities. And if that sort of organization doesn't exist, you have to provide those entities. I mentioned an example with M-PESA. It is all not mentioned that the amount of effort that M-PESA has put in raising all of the agents who are the ones doing the financial transactions. The way they identify those agents has a particular methodology. The way they actually cultivate them and so on. At this point they have more M-PESA agents in Kenya than there are bank branches in the country. I would say that is the real effort that M-PESA has put in to make M-PESA a success. Not the very simple part which is the SMS based technology.

Just to summarize: I want to mention that I talkled about three projects that we have been doing: Multipoint, some research around mobile-phone interfaces, and Digital Green. And then, three myths of ICT. One is that needs translate necessary to business models. Another is that technology somehow undoes the phenomenon of rich people getting richer and the last that information is a bottleneck.

I would say there is one key lesson in everything that we have done. If you want to succeed with technology in developing countries you want to find a way to partner with an entity that is willing to do the other parts that are not technology.

AFFORDABLE INFRASTRUCTURE FOR EVERYONE, EVERYWHERE

Chair: Prof. Jörg Eberspächer, Technische Universität München

9 From Voice to Broadband Data – Affordable Communications in Emerging Markets

Frank Oehler,
Nokia Siemens Networks, Espoo Finland

In Nokia Siemens Networks, we look for many years on emerging markets with the intention to understand the business dynamics and fundamentals in these markets to make communications happen. Research has shown that communications drives economic growth with reference to the studies done by Leonard Waverman with the London Business School and also studies done by GSMA.

In this presentation I would like to take you on a journey to the key challenges we have analyzed in emerging markets to make ICT usage happen. Basically there are three elements I would like to touch on today, which cover the right technology selection, the ecosystem capability in emerging markets as well as basics about the consumers and their demand. Let me start with the last point:

I will start with a view to the consumer and first of all show you the evolutions of consumers in regards to their productivity increase supported by ICT. Here you see an evolution which happens for any consumer. The very first telecom experience is having a shared phone as it is most likely not your own phone. You share it maybe with your parents, or with your friend. And that is when you really make your first call and get the first experience of telecommunications, fixed or mobile. I think the use case for fixed phones obviously is a shared use case. Then, looking into the evolution, the more you use the phone, the more you reach the point when you actually see that there is a business case for yourself to buy an own phone. In this situation individuals purchase their first phone and subscription and have their first personal voice experience. This we see happening in emerging markets in wide scale happening right now and over the last months: A vast growth in mobile subscriptions driven by affordability. Individuals can afford their own phone, and they buy one as their personal business case works. To move now from voice services to internet services is in our world quite easy. We know what internet is and we know about the

A. Picot and J. Lorenz (eds.), *ICT for the Next Five Billion People: Information and Communication for Sustainable Development*, DOI 10.1007/978-3-642-12225-5_9,
© Springer-Verlag Berlin Heidelberg 2010

benefits of internet and its power to increase our productivity and life efficiency. But people in emerging markets might there have a different or even no experience. They might not even know what internet is. And they might have a big challenge to understand what the value could be coming from internet. And that's why we see a different evolution path in emerging markets compared to developed markets. School kids like for example in Germany do need internet already today to download their homework. Internet usage is familiar to them, they understand the value of internet which is in this case the distribution of documents, they use internet also for private purposes and community services etc. So, they know the value which internet adds to their life. Considering that most of these school kids in developed markets already have a mobile phone, they know also about the benefits of mobility. A move from mobile personal voice services to mobile personal internet services is just covered by the addition of mobility. Whereas in emerging market we believe that we need some intermediate steps. The step from shared voice to personal voice looks the same. But individuals in emerging markets might not know about the value of internet. That's where in the evolution we will see a step in which the value of internet needs to be introduced, and we can assume that no individual would invest in internet access without knowing what it can expect from it. Therefore we will see shared internet access as one option for such intermediate step. There are also other elements where we talked just about, very smart provisioning opportunities to the maybe more high-end devices for higher income segments.

We also thought when first of all understanding what is the way of time when consumers walk what are the devices they use, what are the use cases, how they use the technology. The second thing was to understand what do we really need to understand to make services attractive to people. And we are working in Nokia Siemens Networks with a framework of four elements, which we call Affordability, Access, Competence and Motivation. Focusing on emerging markets we all heard about the challenge of affordability. In our research we are looking into a consumer segment of people who are earning 2 to 3 US $ a day. We know that they are spending something like 5 to 8% of their income for telecommunications and any ICT services. So, we have a very good understanding about what is in the pocket for ICT services, which is about 4 to 5 US $ per month we need to meet when designing solutions. Whatever we do we cannot afford building any solutions which are not affordable at the end of the day for consumers there.

Then, of course, we need to have access, very basic physical access to services. From the consumer's perspective that means: How far do I need to walk till I get a signal on my phone? How far do I have to walk to reach an internet café? From an operator perspective it means much more the need to build coverage to reach more and more people from urban towards rural areas.

We need to look at the competence of the individuals. There have been several comments already today. I have seen in the very morning this figure of 98% of

illiterate people coming from the emerging markets. That is a very shocking statement. It sounds like unbelievable high, though the figure is true. If we have a look
on the global level, we will find that 80% of the global population covers literate
people. Something I think we also need to state is that from emerging markets 65%
of the population is literate. This figure covers basic literacy, people who can read
and write. I think that gives a lot of hope. And there is even more hope: Looking
into a study done in India by Sugata Mitra with the title: Hole-in-the-wall. With his
institute, the NIIT he made literally a hole in the wall towards a slum. They are put
in a PC and just wanted to see what people will do with it. There was a young boy
coming and it took him eight minutes to understand that if he is moving his finger
around this mouse pad, that the pointer is moving. It took him another eight minutes
and he understood if he touching it some actions happen. And it took him another
ten minutes and he was browsing in internet. In order to avoid the potential case
that this boy had some internet experience already, exactly the same experiment has
been repeated over years and as a result they have found out that poor and not
literate people can even outperform literate people which have done PC courses.
There is no reason to doubt that literacy as such it is a serious challenge in our
world but on the other side I would like to share that there is lots of hope and lots of
opportunity. These are things we have taken into account in our studies.

Finally I think the most important element to cover is the motivation. What is really
motivating a consumer to use a service? The key to look at is the value provided by
the service. In the course of today we heard a lot about e-health, e-education, which
are all relevant and important topics. I'd like to touch basic entertainment to provide another perspective of value add to life. We all want to help people in their life
and from our perspective health topics are addressing this most, and we touched
base on some of these applications in earlier presentations already. Entertainment
services also create value, of a different kind, but still value for financially constrained people in emerging markets. It is this piece of joy in life which is giving
lots of hope and power to people. At least I made the experience traveling to Tan-
zania talking to the taxi driver which was a 23 year old boy about Franz Becken-
bauer, the famous German football player. This guy knew the name and he knew
many more players and clubs and historic moments of football history. So, he
wanted just to talk about football. That made his life, that's what he talks with his
friends about and that's what he watches with his friends in TV. As much as
eHealth, eEducation and other eServices are, we tend to underestimate the importance of entertainment and the positive impact of entertainment in regards to happi-
ness and joyfulness.

So, if you are looking at these four elements we need to achieve if we want to make
services successful. It is not just affordability. It is not just access. It is not just the
competence or the motivation. We all need to get it together. And I can tell from my
perspective even I have the money and I have a DSL connection at my home. And
I am PC literate and I can read and write. I am not motivated to use facebook. Face-

book doesn't create value for me. All these four elements need to fit in. The same is we need to understand much more the people who are living in the emerging markets. What makes this motivation for them? And what are the great capabilities and competence required for them to use services?

Professor Marsden made one important statement earlier today, which was that we need to consider and reuse elements, technologies, practices, etc. which are in regular use already today. That is so important because when we use already what is on ground, we most likely use something which is in any form already affordable. Something people have access to already. Something they know about the use cases they have to competence with the technology and the service. And they have an idea about what the value out of this service could be. We can benefit a lot when we are looking to things which are around already.

Finally, I would like to show you some of the results from our research and studies about consumers and what is really important for them. It is a bit different than we believe. In the very early morning we heard about this point like financially constrained people are also time constraint. Something we tend to underestimate, but important as we can help individuals to save money and time, which really makes difference in life. If we are looking to the customer experience – and this is actually a point that has been confirmed by our studies and our research -, if we are looking at services and how these services need to look like, how they need to be designed. They need to be locally relevant. There was this statement that content is enough there. I would state if you really look at the local requirements and the local needs of people, there is still a gap we need to fill. We need to proof to people the immediate value. They need to understand that they have a couple of cents or dollars in their pocket. When they spend something and it is just five or ten cents which is not a lot for us but for them it is an investment. They need to understand the return they are getting for their investments. Otherwise they will not spend it.

Of course it needs to be affordable and also services we are offering need to be offered in increments which are affordable for them. It is no reason that we sell a service for five dollars. We really need to sell it in small increments that financially constrained consumers can consume it. As example, a flatrate of USD 30 per month for a broadband connection is not affordable for the earlier mentioned consumer segment which spends about USD 5 per month for ICT. But if you break down this flatfee into hourly prices, things change. You can have the broadband connection on your mobile phone for this 1h, for maybe some cent. As a result we have designed a solution with which you can just instantly subscribe for one hour, for half an hour or some minutes. This is a kind of solutions will help to support a break through.

Finally, I would like to say one thing that is very often underestimated is how extremely brand aware consumers are in emerging markets. The players who have a very strong brand have a very great chance to succeed best because brand is associated with quality and if you invest in something where you believe that the return is the best. We have made studies where consumers have been interviewed and they said very clearly, they would even pay more for a better quality of a network that they can insure that the call goes through. So, it is not necessary that the operator with the best price would gain the most subscribers. It is actually the fact that more and more consumers are valuing the quality of the network much higher.

Basically from a consumer perspective this statement from Pyramid Research summarizes it well: Access technology is becoming increasingly irrelevant for the consumer. I think that is a very important statement for all the discussions about the right technology, about WiMAX, LTE, 3G, HSPA, or what it really shall be. From the consumer perspective, it doesn't really matter. And the more in-transparent it is the better it is.

I would like to look at the technology selection and looking at two scales. One is the geographic spread of population within countries of the emerging markets and the other one is by the income segment. In most of the countries in emerging markets we all know that the basic fixed infrastructure is very thin. In some countries we still see a kind of copper network which is used and upgraded to DSL services or even some fiber to the home. This occurs mainly in urban areas where the highest demand for broadband is, based on the fact that in urban areas we find most of the people who can afford broadband connectivity.

From these fixed broadband examples the price in Africa can be tremendous. I remember when we opened our office in Ethiopia in 2005 we paid something like 2000 € a month for 128 Kilobit DSL connection which was extremely unstable. There are many factors why these prices are that high, which I don't want to touch here.

When we are looking at wireless technologies with higher broadband activity we can move a bit further out of the urban center, we can move a bit further to suburban areas. But normally the business cases for operators are not profitable to move far into rural areas. With the wireless technologies we have a chance still with a fair amount of money to move a bit more out into the suburban area, to cover more people who can afford to subscribe to some wireless broadband connectivity here. But still if we are looking at the rest of the country from the connectivity point of view it looks fairly thin there. We have been taking this as a starting point to understand what we can do to go further out and connect more people. What are the different business models? What is a use case if you look at the leverage of technology which is in place already in urban and suburban areas and also to look at elements and connectivity models which we can be used for rural areas? The shared internet access has been presented beforehand already is definitely is one of the opportunities.

To go further we wanted to reach even lower income segments. There we think about, let's use what is existing already. The GSM networks have been rolled out in the past very strongly. We see a tremendous growth of GSM subscribers. GSM supports SMS and especially USSD in all available handsets. Especially for USSD we see a kind of renaissance as it is a great extension in regards of data services. It is fair enough to say we are no longer talking about internet services but we are talking about very interesting aspects of having and sharing information with people on this low income level.

Quite often I get the question about: What is the technology we will find in emerging markets? Well, there is no answer like that. There will be a multitude of technologies around. There will be 2G, 3G networks. Here and there we will find maybe some WiMAX technologies, in the future LTE. I think the question we should ask is: Who is using what technology? By understanding 'who' is using which technology we get fairly quickly to the answer on how to design an application. This goes strongly in the direction of applications when we for example see shop owners with a GSM phone communicating with entrepreneurs, who have a phone and PC and then finally an enterprise with a full fledged ICT background. So, we have an application which is on the one side IP based, the connectivity between servers is running in IP. However, there is an SMS based interface towards the shop owner. I think the key is really not to say I am building an application which based on the assumptions of available bandwidth for connectivity. So, as minimum I need this 128 Kbit/sec or something comparable like that to make my application work. It is much more about the point that we think about we want to address this service to consumers which have an income level which is something like 2 to 3 dollars a day. So, these people might have a GSM phone. We need to build an application which is communicating with the GSM phone. Then, there is a dealer between the enterprise and the shop owner, an entrepreneur. And this entrepreneur has a PC not always being online but maybe just sometimes. We need to build applications which can work offline, which can be connected to some other services just for a couple of hours or minutes per day. And then of course we have the enterprises with an ICT environment we know best from the developed markets. At the end of the day it is really about to adapt the application towards the value chain and towards the use cases of different players within the value chain to leverage the installed base best.

I also wanted to look at one aspect which is about rural connectivity. Basically I see three core elements which are contributing to cost of building a mobile networks of emerging markets which we call tower, power and backhaul. Tower being the cost of the site construction, covering all the design, management and construction work you need to do. Power being the cost of site electrification. And the backhauling means collecting the site to the rest of the network. By managing the cost of these three elements, we see a great opportunity that we can rollout network coverage in more rural areas. This will be possible as it makes the business case more attractive for operators also to serve people who only will spend a couple of dollars per month in rural, meaning not easy to reach areas.

There are a couple of solutions which I do not want to run through here in detail which are discussed in the industry already today. That is something where we see that technology and state of our technology is requested also in emerging markets to really help to get the cost down. There are lots of features, lots of inventions for GSM networks which are doubling or quadrupling capacity and which are by that getting the cost down for the operators, who are then enabled to forward those cost benefits to the subscribers.

I would like to turn attention to one solution which we have been deciding especially for the demand of emerging markets which we call Nokia Siemens Networks Village Connection. The solution on the one side leverages products from our existing portfolio to reach the important effect of scale in regards to affordability. On the other side we are looking in components like standard PCs which we are in telecommunications not normal at all. On the PC runs special software which is emulating a complete GSM network. So with just one site, Village Connection, it is possible to build GSM coverage in a village at fairly low cost. The calls are routed locally within the village. To connect the village to the rest of the network, a transparent IP connection is required. The innovative part here is that no clock signal towards the rest of the network is required for synchronization. This technical detail enables us to use every available IP transport technology to connect our site and has an impact on the site construction cost as for example no high tower might be required.

There are innovations happening. There are lots of concepts in the industry under discussion. It is just a matter of getting them done on the ground, in emerging markets. It is just a matter of jumping out of the comfort zone and designing and testing such concepts in the field.

Now after having had a look into the technology side of ICT in emerging markets, let's move into the challenge of ecosystem capability in these markets. Very elementary here is to understand the development or evolution service providers are running through. This will help to identify areas we have to address use to reduce costs. In addition this analysis helps to identify areas which are creating an opportunity for us to move in.

I present now a very simple value chain consisting of a network operator and a consumer. The consumer is creating the content, which is basic voice content, in the language the other consumer understands. We can assume that the content two parties in a voice call share is relevant for each other. So in theoretical analysis, the voice call covers user generated content in individual language with personalized features, such as the individual voice, sound etc. This is the killer application voice and it will fly many years in the future as well.

How can we help the consumers and the network operators to cut cost down? We can help of course the operator to run efficient network operations. We can help the consumer by looking into three dimensions: Getting the cost of the entry device down, getting the cost of the service down and getting the tax down. On top of this we will need to understand the consumer pattern in regards to how much the consumer is willing to spend, where to spend and when to spend. The more we know, the more we can gain. There is a very interesting way from another industry which is the household and food sector. I talk about Hindustan Unilever Limited. It is very amazing and motivating how HUL is dealing with consumer insight in this financially constrained segment and how they are translating insight to sell high quality products in very small increments to people. How they distribute it. They know exactly when the pay day is and then they are with a van there and selling these products. Once you understand these patterns you have a great chance to leverage from that also for telecom business.

Looking further in the evolution in emerging markets something will happen differently to developed markets. We have seen the example of village phones. There is an opportunity in the value chain between the operator and the consumer, which brakes up this relationship by having an entrepreneur in-between. That is something unique for emerging markets. The value of this additional chain element is the distribution channel. The value the entrepreneur is creating is the brake down of the cost of connectivity in form of a wholesale service from the network operator into small increments, that it is consumable from the consumer side. So basically this entrepreneur buys a phone, monthly subscription etc. and offers his phone to village members and any 3rd persons as a service, which he gets paid for.

In the earlier presented solution Village Connection we also open the opportunity to execute a different business model. In the case of Village Connection the business model is a franchise mode of operation.

Now we want to establish an efficient way to offer services other than voice. We need content. Operators decide that they get some dedicated SMS based services, something we see in almost every network. Operators start with collecting and aggregating the content themselves. The more variety comes with the services, operators are more and more confronted with a complex environment and headcount consuming activities to select, manage and market these services. The next step is outsourcing. That is basically when we see the content creation and aggregation side in the value chain developing. There are some external service providers which are basically providing the content. There are others which are taking and aggregating this content and building nice applications around that. And there are network operators who finally offer that the content as part of their portfolio. If we look into a kind of revenue share, the operator gains in most of the cases about 70%, the aggregator about 20% and the content creator about 10% of the whole revenue. Such revenue share will change to the benefit of content creators and aggregators in the future. The increase of the value chain is not necessarily an evolution which will help to cut down cost. However, due to the increase of content demand, increase of applications and services, the increase of complexity in the ecosystem will open the opportunity for content aggregators and brokers in emerging markets. Such positions are existing already today, however fairly rare.

In Nokia Siemens Networks we have explored this part of the value chain in several projects in emerging markets, on ground. I would like to take a project from China as example, which we call eCommerce. The intention of the project was to improve the efficiency of a supply chain in China between urban and rural areas. To start with, we interviewed rural population in regards to their purchasing pattern. As a feedback we heard that people visit about twice a month the urban areas to purchase goods they don't get in the rural area. We interviewed rural shop owners and asked why they are not increasing their portfolio. The financial, logistical and storage situation however limits the capabilities of the rural shop owners. We asked also urban shop owners why they don't intend to open shops in the rural areas. However the financial situation and a bad performance of a business case were the hindering reasons there. What we did then was the implementation of a catalogue sales platform. The urban shops could offer their goods in this web based catalogue, we equipped rural shop owners with PCs and connectivity and trained them to offer goods with the PC. As a result after a period of 3 to 4 months we found out that the rural shop owners managed the same revenue with the online catalogue as the do with their existing shop. The urban shop owners could more efficiency deliver and plan and the rural population could reduce traveling to the urban areas. In regards to the value chain described beforehand, we looked in our case into the content creation by the urban shop owners, content aggregation by the catalogue sales platform and content distribution by the rural shop owners. As a clear result, ICT is driving efficiency and supporting businesses to work more efficiently. We have done further projects to analyze the value chain more and are about to identify opportunities for service providers in this area.

To summarize, affordability goes far beyond technology.

- The understanding to impact cost starts with consumer understanding. We need to translate such consumer understanding in use cases and solutions.
- Technology might not in all areas play a role. We have identified that often business models and the consideration of an entrepreneur help to adopt services to the consumer pattern and therefore help to address affordability. The right mix of technology, business model and concrete use case will create success.
- Another important point to add in this summary is the reuse of existing technologies and installed base. There is a great opportunity to build solutions on this installed base, something we tend to underestimate.
- The ecosystem in emerging markets is fairly under-developed. Filling the gaps in the system isn't a job to be done by one entity. A multi-stakeholder engagement

is required, we need partnerships and inter-sector cooperation's to get more projects done, to move projects into scale and create an impact.

- We need patience to get there. In a world defined by profits and growth, new business opportunities are evaluated against short return on investment. However, as the foundation in emerging markets is missing, the return might come later. Often business proposals therefore are rejected. I therefore call for breath and patience as this being a key order in case we are seriously interested in doing business in emerging markets.

10 Leapfrogging ICT with Cloud Computing in Emerging Countries

Dr. Hagen Wenzek,
Corporate Strategy, IBM, Armonk, NY

Success in emerging countries

To understand the impact of conducting business in emerging countries, just analyze the balance sheet of global corporations such as IBM. One might have observed with interest how positive the results were that IBM has reported for the first quarter of 2009 and the quarter before in these difficult economic times. Careful listeners of the remarks during the earnings presentation will have heard where a lot of the profits came from: a successful business design in emerging countries. This provides a foundation for sustainable success to withstand the storm in the developed world. This article describes some core principles to generate very profitable revenue in the developing world.

Muhamad Yunus in his book "Creating a World Without Poverty" laid out a framework of three basic interventions to bring people out of poverty. Firstly, it is about establishing marketplaces of social businesses; secondly, about extending the reach of social services, and thirdly establishing advanced information and communications technology (ICT). The focus discussed herein is on the impact of ICT on social businesses, social services and on innovation in general. Specifically it is on cloud computing and its contribution, explained through a couple of examples that are already happening and partly generate revenue.

Various different stories about the success of ICT in emerging countries have been told and even though they were spotty one can read a lot progress in them. Over the past decade a foundation of connectivity has been implemented around the world. Kazi Islam of Grameen Solutions told the story of Bangladesh where already a third of the population has mobile phones. Even though one might point to the two thirds that are not covered, one should not forget that we are talking about one of the poorest countries in the world. A third of the population can now use mobile phones to do formerly unimaginable things: holding spontaneous conversations with distant partners, accessing information via the Internet and reaching a multitude of other services. This is a development that is spreading across the world, connecting people globally and locally.

A. Picot and J. Lorenz (eds.), *ICT for the Next Five Billion People: Information and Communication for Sustainable Development*, DOI 10.1007/978-3-642-12225-5_10,
© Springer-Verlag Berlin Heidelberg 2010

On the other side we are observing a much broader basis for social businesses. Microfinance has been discussed many times and one finds a solid foundation of supply for micro credit, bringing a critical element of a functioning economy to the developing world. But is it so important to provide credit to the poor? With a brief look at the recession in the developed world the impact of a dried up credit market becomes apparent. Something taken for granted in a well run economy is suddenly missing. As governments spend billions to re-establish the financial markets the same basic element is foundational when establishing a broad based economy in any emerging country, too.

There is already a base for microfinance serving some 150-200 million accounts globally. We have heard about many islands of excellence, great spot examples that are scattered around the world. We have learnt about solutions in Africa, in India, in Latin America and so forth. These are all purpose built, point solutions and they earned laudable success.

The problem of scalability

However, though basically connected, real integration is missing. Therefore, the solutions can hardly be replicated or reused in other places, leading to an enormous supply-demand gap. This can be exemplified by looking again at microfinance, where even though we have are coming close to 200 millions of accounts already served, the actual demand is six fold. So clearly people really are asking for that kind of service. But we have so far no means to serve them all.

In my opinion 'business as usual', i.e. throwing more people and more datacenters at the problem, putting in more customized and proprietary solutions in the market is not the way to go. Business as usual cannot solve the question of scalability. As one example, the usual approach to serve such a demand from the IT side is to build a local datacenter. However, is it such a great idea to build a datacenter somewhere down in Africa, just because you want to run the backend process for a local business service?

Customized solutions are also solutions that are hard to develop. They take time. They take local expertise for IT. What makes more sense it to support more innovative solutions, more entrepreneurship, enable people to solve their local social and economical problems and not for them to become an IT expert.

Cloud Computing as a disruption

What I have been looking for now for the past year at IBM is a new way to handle IT. It is called cloud computing and it has created a lot of hype around the world. Once you start to dig deep into it and understand what is real behind cloud computing, you find it to be a very disruptive play. Weather when starting a brand new implementation of IT, or even when migrating off existing infrastructures.

What is cloud computing? It is a way to create an infrastructure that is highly dynamic while at the same time being very standardized. To make it this way, it uses a lot of standard components for maximum re-use. As a result a lot can be automated so that the whole cycle of maintaining a solution, of deploying a service and upgrading it now operates practically hands free while it before involved keeping many people busy. Last but not least cloud computing implies a very centralized service delivery. Vast numbers of IT users are being served "through the cloud": somewhere out there. This cloud resides in highly efficient datacenters that are located in optimal locations where re-usable energy, low temperatures for easier cooling of machinery and good network connections are plentiful. The cloud provides the infrastructure to deliver so called "cloud services": IT is being regarded only as a service, something where you do not have to bother about its hardware configuration or operating system. Cloud services enable the consumption of IT as a service. IT actually becomes accessible, affordable and very scalable.

Cloud difference 1: Speed of availability

How are the results different? I was struck when for the first time I observed the deployment of a virtual machine – a server somewhere in the cloud. It was already tailored for my needs as a fictitious developer being in China in a technology center who wanted to do some development. Usually, if you need a server to develop on, you would conduct an acquisition: filling out a requisition request form, submitting it and after a couple of weeks or months the equipment gets shipped, somebody sets it up and you are proud to have a piece of machinery under your desk. With cloud computing you are getting the same result in some 90 seconds – just without the box in the office. When before you had to wait for weeks and months it is now available in less than two minutes. A truly disruptive difference.

Cloud difference 3: Service catalogs

A second very different result of cloud computing can very easily be understood when examining the changing business relationship between a user and a provider. Today, most companies still decide to develop, run and maintain their own IT. Some outsource that IT department to a service provider, but they typically just ask to have the same "mess" managed for less. All this requires lengthy contract negotiations, long-term commitments, renewals etc. Adapting to changes is often cumbersome and not in synch with the business requirements. With cloud computing you go on the Internet and make your choice from a services catalog, something that might resemble a Chinese menu at a restaurant: you select M24 with A17 and a couple of spring rolls and pay as you go. The same can be found in cloud computing: You select the application, size of project, performance and only as long as you actually use the service you pay for it. The moment you have done what you wanted to do you automatically stop paying. There is no need to do anything else. This approach is a very different way to select a service, negotiate a contract and pay for the use. Therefore any experimentation can happen much swifter.

Cloud difference 3: Diminished up-front investment

Last but not least there is another very important difference: the upfront investment shrinks dramatically. When looking at a typical setup of IT infrastructure in small or large enterprises you are running into thousands or millions of dollars to start using IT seriously orzo. A datacenter gets built, servers ordered, operating systems installed, software customized and everything kept up to date. At IBM we calculated that typically you spend easily between one and ten dollars per machine hour for a normal server when adding all the labor costs involved. So, every hour a couple of dollars for each server installed.

With cloud computing you get the same result, but no machine is actually shipped, no software installed. There is practically no up front investment left. This has been taken care off by somebody else who operates the cloud and only delivers cloud services for just a couple of cents per hour. One classical sales price for virtual machines nowadays is 10 cents per server including storage and normal memory and the maintenance is included, because it is automated.

Making the difference: Three project examples

How is all of that cloud computing being used? And why is that so important for emerging countries? Let me give you three examples.

Example 1: Cloud-enabled micro financing

The first one we have talked about a number of times already, but now I am looking at the backend of that process: The operations of a microfinance institution. Why are they today not able to fulfill the demand of the market? The lack of scalability is caused by a large number of manual process steps. E.g. loan officers are serving their clients using a normal notebook; not a laptop, but a paper notebook. Here they document and trace all the loans in their area. A spreadsheet will at some point of time somehow be assembled at the bank and consolidated into the balance sheet. However, the result is not really something that other institutions might fully trust as it is based on many failure prone manual steps. Secondly the balance sheet it is hardly auditable. So, anybody who wants to invest into the microfinance institution to provide the essential liquidity for more loans is often stepping back saying: "it is maybe not such a good idea; I don't trust the paper work that is going on".

To overcome these problems we developed a micro financing process hub. A solution that automates all backend processes, starting with the loan officer now using a smart phone to track accounts. The phone will automatically transmit any status change to the branch office so the staff people can work with it. All is accessed through the cloud, like our cloud center in Dublin, Ireland, where the micro processing hub for Latin America is hosted. A couple of millions of accounts in Latin America might therefore be processed from across the world in Europe. But if I had not told you, you would have not known. And while today it might still be in Dublin, next week it could be somewhere else on the planet. It really does not matter.

We automate that backend, decrease investments by the enterprises and therefore make sure that the startup costs for micro financing companies are being reduced dramatically. So, the barriers to entry are being lowered or even removed altogether. The hub naturally comes with a robust compliance in place, too. Furthermore, it acts as a platform to leverage the current business design and seamlessly expand the portfolio with new services. New applications that support other financial services in the emerging countries quickly become part of the platform and can be used at an incremental fee. Together with a few partners, we are working to offer such a more comprehensive cloud service platform to interested companies.

Example 2: Cloud-enabled innovation infrastructures

The second example that I briefly alluded to is a center that we established in the City of Wuxi, China, a special development zone. Even though the scope of this conference is about the whole bottom of the pyramid, there is a huge market that starts from the lower top of the pyramid and extends down towards the bottom, often making it a better business case to justify. One market covered by that part of the pyramid is China where we enable local entrepreneurs. A cloud computing center was deployed by the City of Wuxi. From this innovation center the city provides an infrastructure for local entrepreneurs to write new applications. These are applications for local needs, maybe global needs. It depends on the ideas and the innovation and the business model of that entrepreneur. What an entrepreneur does is to go to the center and start using software packages that normally take a lot of money, a lot of effort, a lot of time to set up. They now can use them on the fly. The whole platform is being provisioned through a very dynamic infrastructure from this center in Wuxi. Respectively the local government is able to incentivize the entrepreneurs to leverage a local infrastructure, be innovative and spark economic growth on a local scale.

This center has become highly efficient allowing them to take the same concept and offer it to other cities in China. By merely adding a couple of extra machines to their normal IT production site they will be able serve many more developers around the county from that one center. Within the other cities the infrastructure and applications appear exactly the same. Entrepreneurs can do the same stuff on the fly, very fast, very cheap. The cost to deliver a service like this instead of doing it the normal way, shipping machines, installing PCs and so forth is just a fraction.

Example 3: Cloud-enabled healthcare education and information processing

My last example goes back to Africa. It touches a couple of the points that we have been discussing: Education on the one side and health on the other side. What is happening there is a collaboration with a number of Eastern African based universities and their health departments. IBM is helping them to collaborate to serve local communities with education services from an open virtualized learning management. Distant learning or virtual training is critical for the success of an education program. The project is about getting content to the people, create trust to collaborate in learning and make it much cheaper for everybody who is using a PC, since the desktops are now served by a cloud center and not being installed locally. This way many more people get a much better medical education: health care professionals as well as local citizens. These new programs for education are provided by the universities at the local level, coordinated by the HEALTH alliance.

In addition, health care services are being provided through the same organization and the same solution via mobile health care services. Mobile phones are the front end for ICT for many people: The health worker as well as the end user. One application that is supported from this program is to use medical records in an open fashion to automate the backend processing of the medical record systems and to enable healthcare analytics. This program makes both research and treatment much more effective, by basing it on actual evidence generated from the data that is on the cloud.

Right now the IT services are delivered from am IBM cloud center in Pretoria to these seven Eastern African universities. That center acts as the base for the infrastructure running an instance of the open education system SAKAI, an open source system that is starting to spread really around the world to provide a foundation for a wide variety of education services. That platform is being extended primarily around education, but also around these domain specific services such as health care. At some point in time this IBM hosted solution should be replaced by a local cloud center somewhere in the Eastern part of Africa to overcome potential connectivity issues. But even though I am telling the story it does not really matter from the delivery standpoint. More importantly, the alliance will have done some heavy standardization, be open and only deliver a service, not machines – thus fulfilling the principles of cloud computing.

This last example of cloud computing is one that I think shows the impact once you stop thinking about the business-as-usual approach towards IT in emerging countries. Yes, we have a foundation of connectivity and social businesses that already work pretty well. A lot of social businesses are supplying critical goods and services. The situation is good, but not scalable to match the full demand. So we see a great opportunity with some challenges.

Summary

By taking a cloud approach to tackle the issue of scalability and make it an opportunity you are looking at standardizing, centralizing and simplifying IT, because it really dramatically improves performance. After closing the scalability gap one will see that IT has fully transformed into a pure service. One that is much more accessible, affordable, and scalable.

The result of the efforts will be sustainable. Local entrepreneurs will get the support for their ideas through the provisioning of an innovation infrastructure. Credit will be available to fund the business plans. And lastly healthcare education will be available even in regions that are just hard to reach.

Overall I think with clouds we have the opportunity to just leapfrog the business-as-usual approach in the emerging countries to help more people get out of poverty faster and do it in a way that is compatible with business requirements.

11 PANEL DISCUSSION
ICT and Economics Development in Underserved Markets – A Chicken and Egg Problem

Chair: Prof. Dr. Arnold Picot,
Ludwig-Maximilians-Universität, Munich

Participants:

Anriette Esterhuysen, Association for Progressive Communications, Melville, South Africa
Dr. Arnulf Heuermann, Detecon International, Bonn
Samia Melhem, The Worldbank Group, Washington D.C.
Thorsten Scherf, Federal Ministry for Economic Cooperation and Development (BMZ), Bonn

Prof. Picot:
Good afternoon to everybody. We are approaching the final hour of our conference and I have the pleasure to introduce to you our panel this afternoon. I start with Anriette Esterhuysen. She came from South Africa to this conference, and she is the executive director of the Association for Progressive Communications, APC. This is an international network of more than 40 civil society organisations dedicated to empowering the supporting groups, especially with respect to the strategic use of information and communication technologies. Before, she was an executive director of Asengonet, an internet service provider, a training institution especially dedicated to the development sector in South Africa. She is also a founder of Womensnet in South Africa and has served on the African Technical Advisory Committee of the UN's Economic Commission for African information society initiative. She was also a member of the United Nation's ICT task force. So, I think she is best equipped for our discussion this afternoon.

Next to her is placed Dr. Arnulf Heuermann. He is a managing partner of Detecon. Detecon International is a consulting group based here in Germany which is among other things well known for worldwide consulting services for ICT infrastructures. He has gathered experiences in more than 35 countries over the years and has held responsible positions with respect to strategy, marketing, management and international sales. His major substantive foci in his work are telecommunication regulations, merger and acquisition, liberalization and privatizing activities as well as market and corporate strategy. Before he joined the Detecon group he worked for a couple of years for the WIK, Institute for Infrastructure and Communication Services, a quite well known state owned research institute here in Germany and gained a doctor degree in economics from the university of Bonn.

To my left I can welcome Ms. Samia Melhem. Samia is a senior operations officer in the global ICT policy division of the Worldbank Group in Washington and she is a member of the e-government practice group there. She chairs the e-development thematic group, whose mission is knowledge sharing amongst ICT practitioners, also a very important subject for our discussion today. She serves currently on special groups with regard to ICT policies and ICT in the public sector in the Worldbank and addresses quite a few other related subjects. I might also mention that she has held several positions in Africa, Middle East and Eastern and Central Europe. So, she knows various regions of our world with respect to our subject. She has gained among other degrees a degree in electrical engineering, in computer science and an MBA.

Last but not least I am very pleased to welcome Thorsten Scherf. Thorsten is a member of the Federal Ministry for Economic Cooperation and Development in Bonn. There he serves as an advisor on information and telecommunication technologies for development which is really our subject. Before he worked and did research at the University of Marburg focussing on regulatory matters for providing universal access to telecommunications in rural areas of developing countries. Among other countries he gained practical experience and did research in Peru, Bolivia and Uganda and he gained a doctorate degree at the University of Marburg. But before that he served for a couple of years at the Arcor AG, a German fixed net provider at the time. There he was charged with regulatory affairs questions. Thank you very much for joining us.

Ladies and gentlemen, we start our panel discussion and I have asked each of the panellists to bring forward some key points that they think are most prominent and most important in order to understand the topic of our panel which is called "ICT and Economic Development in Underserved Markets – A Chicken and Egg Problem". Before I invite everybody to present his or her major messages I would like our technical staff to present once more one of the slides that Joachim von Braun presented this morning (see figure 4 on page 9).

If you look at this slide you can ask yourself now 'Where are we going?' Is the gap between low income and the middle-income regions closing or widening? If you look over the years it seems to be widening quite a bit. This might be a concern. We want to discuss whether it would be possible to close that gap by either infusing more technology in a holistic way – we have learnt from Dr. Toyama today that a comprehensive holistic view is very important – or are other factors at play so that this gap perhaps will be difficult to overcome? So, what is the interplay between the economic development in terms on the level of income that is displayed in this slide and the diffusion of technology, especially ICT technology? I think this could be just a visual starting point which really describes that we are not talking about an academic question but a real world issue that is I think of some relevance for us here in the so called developed countries, but even more for the entire world. Now I would like you to bring up your comments. May I ask Anriette Esterhuysen to start?

Ms. Esterhuysen:
I want to start by saying that I are really enjoyed the inputs thus far and found them very interesting. In response to the chicken-egg question: I don't think it is a chicken or an egg problem. What we really need is a chicken farmer because if we have a competent chicken farmer we will have lots of chickens and lots of eggs. And I would like to think that chicken farmer will employing people from the local community, and using renewable energy resources, and that the chickens are free to roam about and scratch in the way that chickens like. If the chickens are happy and healthy, then the eggs will taste better and contribute to a sustainable community development initiative that employs lots of local people.

So, I really think that the answer to the question does not lie in whether it is the technology, of the information and communications that comes first. If you are talking about creating sustainable economic development you need a holistic approach. You need more capacity development. You need financing. Good governance becomes very important.

The power of information and communications is that it is an enabler for all the other building block that we know is needed to generate sustainable development. We need to consider how this enabling role can be emphasized and take place more effectively. Building the necessary infrastructure at local is particularly important. By infrastructure I include a range of different kinds of infrastructure, from technical, telecoms and internet infrastructure to institutional capacity, business development opportunities, support and training, etc. You need ongoing human and institutional capacity development.

In closing I want to reflect somewhat critically on initiatives intended to tap untapped markets, and reach the so-called bottom of the pyramid. It is not that I believe such initiatives are not important. I think they are important, and they can change the mindset of the private sector and of government institutions that have tended to ignore the potential of investing in poor communities. But I think to believe or to imply that tapping these markets will create development is problematic. I have no doubt that investing in product and service development for previously untapped markets will add value. It can create jobs, and opportunities, and introduce new ways of delivering important services. But will this produce sustainable development? And, what about the impact of people who already have very limited financial resources being caught in supply driven spending, resulting in them saving less, and perhaps spending less on essentials?

The reason why I think it is dangerous to make the claim engaging previously untapped markets can create development is I have seen the damage that such 'overstating' can do. I have been working in ICT4D since the 1980s and I believe firmly in the power of ICTs to support development. But I have first-hand experienced of the damage that, for example, the notion of "leap frogging" can do. Perpetuating the notion that the kind of investment in development that has taken centuries in the rich parts of the world (e.g. through building strong public institutions) is not necessary in developing countries can lead to unwise investment decisions, by governments, and by the donor community. Of course investment in development today does not necessarily have to take place in the same way that it has in the past, but the fundamental building blocks for development such as human capacity, freedom from conflict, good governance, food security, health and safety and greater social equality should not be neglected.

The assumption that ICTs in underserved markets, like the assumptions prevalent a few years ago about ICT4D being somehow a "magic bullet", or the notion that there are "killer applications" that can make it possible to bypass some of the long hard slog that it generally takes to build more equitable social and economic development, are dangerous. These assumptions actually discredited ICT4D. Perhaps this is the reason that so many donor agencies are no longer emphasising ICT4D to the extent that I believe that they should be. It is almost as if they are somewhat embarrassed by having leapt onto the leapfrogging bandwagon and now, instead of strengthening their investment in ICT4D by ensuring that it is holistic and sustainable, they are either moving into new directions, or mainstreaming ICT4D to the extent of neglecting it.

ICTs in underserved markets are a positive and powerful trend, but it is not enough.

Prof. Picot:
Thank you very much. Let us turn to Dr. Arnulf Heuermann and listen to what he is going to tell us.

Dr. Heuermann:
You asked us to prepare five theses, – I would like to mention four. You showed us this nice picture of colleague Prof. von Braun on digital divide and the development of fixed, mobile and internet. If you look at it right you will see – and that is the first thesis – there is a bottleneck resource that is called fixed network. The lacking investment into fixed broadband access and broadband backbone networks may widen the digital divide in future, because the real success story is only mobile voice communication. In fact that is the only area where we have a shrinking of the digital divide, and where we have a real market success in developing countries. The problem might be the demand is going on for broadband access. We are discussing in our countries about fibre to the home and fibre investments because we think that the access bandwidth supply would be too small for future web2.0, video demand etc. If you look at countries like Ghana than this country has a fixed

access network for about 150.000 people but 4.5 million that makes 30 times that number of subscribers are mobile. So, the disbalance of fixed and mobile networks is extremely strong. Fixed operators typically are loss making in emerging markets and regulation that regulates wholesale products, (e.g. interconnection rates between fixed and mobile) are in many cases set to the detriment of the fixed network leading to additional losses in the fixed business. Most likely that might mean we will have a lack of investment into the backbone or many mobile operators will build up their own backbone infrastructure. But in many countries they don't share this infrastructure leading to inefficient double investments.

What does it mean? Incumbents might survive in these countries if they invest into wholesale products. That means serving the mobile operators and mobile access operators with backbone and aggregation network products. Or they will be become takeover candidates and that is what most of them are. The takeover will most likely be done by the mobile operators because we have this tendency to fix mobile convergence of course. That might be the outcome.

The second thesis is then on the impact of the financial crisis. Who are these successful mobile operators I was talking about? If we look at Africa then the number 1 is Zain, number 2 is MTN, number 3 is Etisalat, number 4 is France Telecom, number 5 Vodafone. What we see is these are the regional or global players. Most of them are not of African or regional origin. These companies are all private and many experts do agree that the bulk of investments for the success in ICT comes from private investments. Although their role is very important, public finance and development aid plays a minor role in ICT financing. The major investment, the billions of Euros that are required for the infrastructure rollout come from private investments. So, what does the financial crisis mean for these investments? The good news is – most of these businesses are cash machines or "money printing" business. In many countries, in particular in emerging countries, more than 90% of the business is prepaid regardless of regular fixed or mobile. That means you print money, in other words "rights" for people to have future telephone calls or whatever. That also means ongoing operations are not tackled and also local demand is not decreasing much. At the moment the core business of telecommunication is not very much impacted by the financial crisis. The bad news is that the discount rate used to calculate the value of the company has increased strongly. On the one hand side it means that the price to buy into a communication company has decreased extremely in emerging markets which means that the sellers, mainly the local governments don't want to sell and wait for a better time. And on the other hand side also the big players like the Vodafones and MTNs have difficulties and many of them are listed at the NY stock exchange or somewhere else, where analysts are looking at their debt structure and they cannot invest as much as before. In fact the structural change needed in many of the emerging markets is more or less delayed now. About 15 ICT privatisation projects in Africa alone are now delayed or not contracted etc. This might be a danger for ICT development in the future.

Third thesis is about wireless broadband access and spectrum availability. We saw a lot of nice charts here on the wireless access, on LTE etc. WiMAX for example is one these technologies where we currently say, this is the magic broadband access technology of today. Unfortunately it not really suitable for rural areas because if you want to realize inhouse coverage in tropical areas and you look at the cell size you can realize there then it is 1.5 to 2 km perhaps. That is a technology for agglomeration centers, for a big city. In rural areas it would be much too costly. Each WiMAX base station has to be connected by a broadband cable. And deploying that every one or two kilometers is very costly. How much is the CAPEX requirement for fibre to a home in Germany? 30, 40 billion Euros? I think this was one of the estimates here.

So, we need another spectrum. Three, four, five GHz is a physical barrier to big cell sizes which we need for rural areas. We need a spectrum refarming. That could be in the 900 MHz or in the GSM band or it could be in the 700, 800 MHz band the former analog broadcast spectrum. Like in the US, where it has already been allocated to broadband internet access and maybe the technology is suitable for serving rural areas in emerging countries. This new technology will then come from the US and no more from Europe like GSM which was one of these success technologies before. Most likely the next big ICT investment boom will come from the US.

The final thesis is on universal service policy and social responsibility. In emerging countries we still have these fixed line incumbent which mostly are public operators with universal service obligations. And then we have the bright shiny rich and privately financed mobile operators which in many cases don't have any universal service obligation although de facto they are the universal service providers in the country. And many of them have company policies on corporate social responsibility which includes topics like anticorruption trainings, rural communications, supporting e-learning and e-health initiatives etc. But we must say this is a "good weather" economic attitude. If times will get rough and if you have five or six operators in a country like Ghana one can be sure that there will be a shake-up within three to five years. The question is who is then doing the unprofitable investments in rural areas? Maybe that means we will need a revitalization of universal service policies and we might also look at universal service obligations for mobile operators. Thank you very much.

Prof. Picot:
Thank you very much for these very interesting inputs. Let's move forward now to our next speaker Samia Melhem from the Worldbank Group. Please, go ahead.

Ms. Melhem:
Thank you Prof. Picot and I want to thank the organizers from Münchner Kreis for inviting us today. I had something ready that I completely changed after all the good presentations. I think I want to focus with two specific points. The first point

is the growing contrast that I observed personally between where the private operators are going and where the mobile and the state of readiness in terms of ICT adoption by most developing countries governments. In my view the contrast certainly is growing deeper except for a few countries that we hear a lot about, mostly the countries that are doing quite well in governance, rule of law and fostering a good climate for business and for innovation. Look at us here around the table. We have between Nokia Siemens, Vodafone, Microsoft – you have all emerged yourselves in developing countries culture to understand the consumer and of course with a long term view with prospect of revenues and long-term plans. You understand very well the trends of these consumers, much better actually than do a lot of their governments. And here we are at the Worldbank for the last ten years investing a lot of money in the ICT space, both in the telecom liberalization and participation space which has turned out to be very successful for us. In fact we financed a lot of the many applications you have heard of today such as Voxiva, Manobi, African Virtual University, and several knowledge tools like the ICT Toolkit for task managers and the Telecoms Toolkit. But we have also financed experiments and big projects in e-governments that are not doing as well as they could. We have a large portfolio of e-government and IT applications for public sectory firm of around 7 billion dollars. Often we want to know what is happening with this in terms of change and government transformation and process improvement impact. Some of you here are talking about 2 million to develop an agricultural market place. That is quite cheap and with the outcome and the indicators you are showing these look like very successful projects probably as they are lead by their real champions.

Second point I want to make is that I think there is a big contrast there with the skills that are now going to the private sector, with the top match people that are hired to work on ICT and the skills in the government in the public sector for developing countries. Unfortunately the resistance of governments to outsource a lot if the IT work to the private sector concerns some legitimate ones like maybe loss of privacy or security but others that are really loss of control. I think the presentation on cloud computing was very powerful because the costs are so low today. Why do we have to pay so much for e-procurement systems or a tax system or a customs system when there are solutions that exist there that make sense and that are cheap? Let us use the rest of the money we save to deploy, to rollout, to get out of a pilot phase that hasn't ended for the last ten years. We have been doing pilots here, pilots there but never attaining that critical mass that everybody talks about. For me critical mass is not 15% or 20%, it is 100%, 100% of the population that has access to these services that would eventually impact them and improve their likelihood. For me that is a very important point that we have to keep drilling down and trying to understand.

The other issue that I would like to discuss is about engagement and inclusion: how we can through the success of mobile operators, through the success of indigenous adoption of technology also push for people not being only consumers in this country but being producers? It is very good to have innovation that is coming from Washington or other countries from the North. But how can we push the local people to adopt and to innovate? I think we heard some good stories from South Africa. But how can you push more for it? Because at the end it is the win-win. My specific focus is also: how can we have women be engaged in that innovation culture? Women as you know when you look at the worlds poor there are 60% of the world poor. They constitute the major parts of the illiterate. Yet, they would be one that will benefit the most from having access to information in terms of planning their live, planning the number of children they want to have, curing the diseases they have in their families etc. How can we kind of focus a little bit on that part of the population that really needs help and access to the right information?

Finally I would like to talk amongst us and maybe enlist part of your questions about how can we partner to get a real public private partnership. It is very obvious that not one body can do it all. The governments cannot deliver all this, the private sector cannot. How can we keep this beyond the transition of meeting in these very good events? But do we demobilize ourselves together; put our means together for this project's implementation. I think that is the only recipe for success that we will have. How can we all give what we have learnt in the past and today put the best solutions possible for the developing countries?

Last but not least I think the issue of skills is a very important one because we are creating very good solutions in many of the countries. But there is nobody left to maintain them and then two, three years later because of lack of maintenance, because of the lack of the culture of maintenance and handholding and training people and cleaning up. It is not a glamorous job to maintain IT. Very few people like to do it. Nobody likes to maintain data. So, very good systems that work right in year one last till year five, as the data (not the system) become obsolete. We are talking a lot of data basis that we have financed for taxes, for customs, for population registry. In the beginning they look great and I think we have with us here who probably have a lot of experience in this big ERPs investments that die off not because the system is not good but because nobody has maintained that, because nobody invested in data base maintenance, accuracy, infrastructure maintenance etc. How can we together help developing countries create that culture of IT skills and IT maintenance without which everything that we are talking about today would not be sustainable. This will also help create many jobs and the foundation of a small information society. These are my points.

Prof. Picot:
Thank you very much. We see already that there are a lot of open questions and future challenges on the table already. Now, let's turn to Thorsten Scherf and his view on the subject.

Mr. Scherf:

Thank you, Prof. Picot. Many thanks to the organizers for inviting me and many thanks also for bringing the topic of ICT in developing countries back on the agenda of the German ICT community. I would like to sketch in a few words the approach of the German Development Cooperation in the area of ICT. This should give you an idea how we see the relation between ICT and development. Just as a basic orientation: We are convinced that ICT for development will only be successful to that extend it is demand driven and market based. Development cooperation should focus its contribution to cases of market failure. Additionally it is the intention of German Development Corporation to support the use of state of the art technology in line with local demand and local circumstances in all sectors of development cooperation. This is what we call mainstreaming. In German Development Corporation ICTs are considered an important instrument to advance socio economic development and support countries in the chosen path towards the millennium development goals. Therefore we promote and improve framework conditions and private sector development in the ICT sector as well as ICT transfer and appropriation across various sectors in particular the priority areas of German Development Cooperation. ICTs are seen as means and not as ends in themselves. ICTs are supported by financial cooperation and employed in capacity development programs of technical cooperations, in sectors like governance, education, health, environment or business modernization programs. Since the beginning of 2008 we have in the area of ICT a special focus on supporting sub saharan African partner countries, in generating adequate regulator frameworks in the ICT sector. Those frameworks should insure competition, development, promoting investments and also affordable access to ICTs for poor people and those living in remote and rural areas of developing countries. Currently we are supporting Sierra Leone and Benin in this matter. There are also programs with a focus on IT training and blended learning, fostering business opportunities and virtual networking, and promoting innovative ICT solutions such as free and open source software. Last but not least we recognize the particular importance of public private partnerships in the area of ICT for development. As some of you know better than me, German and European ICT companies are today involved in numerous emerging markets. It is an objective of German development cooperation to mobilize the creativity and innovation of the private sector in the interest of long-term development goals. Early examples include "Africa Drive", a cooperation project together with SAP and Siemens in Southern Africa.

Prof. Picot:

Thank you so much. Ladies and Gentlemen, you have been offered a full menu of issues and questions and also influencing factors that might drive this field and shield the field. I must say that having heard all this I am not quite sure whether we know now whether this is a chicken or egg or both. But that is perhaps not this important. We have learnt that there are lots of factors at play that must be taken into account, that must be addressed, that must be understood. So, I want to discuss on the panel, but I also want to invite you to come up with your questions, your ideas and your critical comments and additions to this very important subject. Yes, there is the first question.

NN:
We are looking for the infrastructure and you said that there are massive investments for examples from the Worldbank. And on the other hand we saw this wonderful little project there going on. I think we should focus in a way to combine these two issues and fund financials directly into that seeing that we have a problem regarding infrastructure. I think Dr. Heuermann can add there something. But there I think is the focus where we should find entrepreneurial models or something like that. Thanks.

Ms. Melhem:
Especially now with the current financial crisis we are increasing the investments in infrastructure in most developing countries. And we just approve the biggest loan ever for Kazachstan on infrastructure build-up. So, we are really trying to help increase the access agenda especially in rural and isolated areas. There are still some places in Africa where constant access to broadband is a monopoly. As you probably know in some countries in Africa, the monthly cost of broadband is ten times the average year's salary. So, it is completely disproportionate compared to Europe, America or Australia. The key for this will be at the same time a combination of competition, more players, more entrants, cheaper services, clever partnership for the mobile operators also who are playing a very important role in that space and as the speaker from Detecon said more intelligent spectrum policies. The problem we have in a lot of the developing countries is the regulators really need to catch up with the technology. And regulation is always behind. So, how can we empower the regulators who most of the time are really and want to increase access and decrease the price but do not know sometimes how to formulate these policies? This is something that really is intensive capacity building. We are trying to work a lot with ITU to train the most regulators based on the latest technologies and the best ways to increase access.

Prof. Picot:
Thank you. Dr. Neumann!

Dr. Neumann, WIK:
I have a question to Arnulf Heuermann. I very much share with you the analysis of the infrastructure of relationship of fixed and mobile networks and the potentially upcoming problem that we might see in the developing world. We see it also in some European countries already upcoming, certain bottlenecks in the fixed net. Where is the solution? Is one idea the solution that we are now going to take the system the other way round? That we tax now mobile operators so that there will be

some financial screen going from the mobile industry to the fixed industry? The opposite way as regulation worked so far. Is that a model or is it not?

Dr. Heuermann:
No. I don't think that we need to subsidize that very much. We should give them the freedom and also the strategic focus to really make a business out of wholesale business. Up to now most of the incumbents just offer some interconnection and termination services to the mobile operators, because they are forced by the regulators. However, business could be much more if they really say: we are the backbone providers; we are the efficient aggregation network constructors. That could be a good business. As you know BT is making nearly 50% of its revenue from wholesale business and nearly 75% of its profit. This could be a model meaning re-focus away from retail business and telephone subscribers.

The second possibility of course is that the fixed operators should jump to the train on broadband connectivity and maybe in this case not so much for private households but at least for the business customers in the developed cities etc. That could also be a lucrative new growth at least.

The third opportunity is pricing. Currently the pricing or costing for termination of these fixed lines or in mobile business usually results in higher prices for terminating a mobile network than a fixed network. This is no truer, really. We did such an analysis in Uganda. Extremely high copper prices over the last years plus shifting of traffic (more traffic is now transported in mobile network in many countries than fixed networks) leads to a different cost structure. There should be a higher termination fee in fixed lines. If you just adapt the LRIC models and allow these companies to set prices accordingly, they might be profitable. On the other hand it could be a good opportunity, too that they are taken over by a strategic investor. These might be the mobile operators but not necessarily. Of course commercial ideas are not the only point why people invest in emerging countries. There are some investors that strategically want to secure resources and therefore invest in other countries. Obviously Governments should be careful with such a partner. But in general it is a good idea that inefficient old incumbents are taken over by others. If you look into countries where companies like Vodafone, France Telecom or others took over or got a new licence they often managed to get 50, 60, 70% of market share from the beginning just because the know the business better, they have a better sales and marketing strategy etc. In a way it might be a good thing to be taken over by someone who knows how the business goes.

Prof. Picot:
Thank you. Would you like to comment on that, Anriette?

Ms. Esterhuysen:
Thank you. Just a few comments in addition to that. I think they already are paying quite high license fees. One challenge is to re-think universal service. I completely

agree with you about the importance of a basic broadband backbone. It would be much more useful for regulators to take the license fees that at the moment are sitting dormant in universal service funds and invest that in creating a good broadband backbone. Because we now know that wireless, with mobile there are so many last miles solutions now. You don't need necessarily a fixed network for the last mile. I also think that regulators and policy makers need to think about the mobile industry. We had before a talk about tower sharing. In my experience in most parts of Africa mobile operators, unless they are forced by regulators, are very reluctant to share infrastructure. This is something that can be forced, that can be compelled by regulators. Pricing as well. So many of the innovative solutions we heard about this morning relied on SMS. The profit margins on SMS or the mobile industry is beyond imagination. It is very expensive. In South Africa you pay about 80 cents for a SMS. I do not know the conversion rate, but it is extremely expensive. So that is something else. Bring down, compel mobile operators to charge less for SMS. Standards and interoperability become increasingly important as more applications and services roll out to end users through mobile. Again, looking at policy and regulatory interventions among that, to keep that as open and ensure open standards and interoperability as much as possible. I think the point that you made about this rely. I don't think it happens everywhere but certainly in many African countries. These governments rely on large operators, who are in fact the new incumbents, to invest in health care, invest in education. I think this is quite fragile and this is really building public sector infrastructure to deliver services over the longer run. We could see these investments actually to be stopping quite suddenly, based on economic or political circumstances. What we also see at the moment is a lot of vertical integration, with mobile operators also waiting to invest in online content and other service provision. This is quite exciting on the short term but on the longer term it could make the environment even less competitive and stay full of innovation.

Prof. Picot:
Thank you. Would you go ahead?

Ms. Terrenghi, Vodafone Research:
Where do you see the opportunity for ICT to enhance the reaching of a certain standard of lifestyle on the one hand, and at the same time maintain cultural identity and cultural differences?

Prof. Marsden's presentation mentioned sustainable design: Do you see any potential interesting partnerships or strategies in the developed market and this field?

Prof. Picot:
Who would like to comment on this?

Ms. Melhem:

I think this is an excellent question. On one hand this common platform we have, somebody was joking that we don't want to get onto facebook, is forcing from kind of standardisation of approach and usage especially for those who speak English. But on the other hand I have seen a lot of very creative innovations and solutions in specific countries. I give you an example. We had officials from Pakistan last week at the Worldbank. They are using mobile phone for banking, for identity like e-ID and for mobile Hajj services. That is something very specific to that particular culture and helping the travellers. We are seeing a lot of very similar cultural heritage applications in some of the projects we are financing in Latin America and the Caribbean. In fact tourism is one of the sectors that is really looking to ICT to refuse to the whole world, natural heritage, touristic sites etc. I am working right now on Vietnam and I am seeing a lot of this translated from Vietnamese to English and other languages which is very specific to the country itself. I think you are going to see at the same time for those we speak the same language, mostly English, a convergence of use. But you are also going to see as typical mass increases in local context, in local culture appearing on ICT.

Prof. Picot:

You would like to add something, Mr. Scherf.

Mr. Scherf:

One of our implementing organizations offers capacity building programmes in the area of free and open source software (FOSS) programming. This enables our partner countries to create software with regard to the local circumstances and local needs. One example is the translation of the Linux operating system in Khmer language and letters in Cambodia which was supported by German development cooperation.

Prof. Picot:

I would like to remind you of Dr. Islam's vision this morning that translation services might be a very important factor in this field because this would enable smaller cultures to guard their proper language and at the same time being able to communicate. Of course, there are many barriers to this project, but still it is a very attractive vision.

Mr. Hellmonds:

I am a former colleague of Samia at the World Bank. Today, I am head of corporate social responsibility at Nokia Siemens Networks. I think it is quite interesting to see that terminology coming up here. One of the things that I am working for is to move basically the social responsibility thinking into the mainstream of the business rather than having it as an add on. So, I am very happy that my colleagues here are moving in that direction. This conference is a good example. A couple of years ago, 15 years ago, I was pushing some of the pilot projects in Africa, especially

Zimbabwe, Mozambique and in Conacry. At that time already I said that I wanted to move away from pilot projects. But even today we are still in many areas engaged in pilot projects. I think if we want to move from a pilot to rollout on a large scale what we really need is not to have it as a CSR project. Sort of like we do good and where we always say, my budget is like this big. It is like a thimble full if you want to take that as a comparison. But if you really want to move big scale it needs to be for profit business because only that really gives you economies of scale. If you look at the mobile expansion in many parts of the developing world and especially in Africa it has been driven not as a do good kind of a thing but it is an expansion driven by business and done with a for-profit motivation. What I am wondering is how can we move to bring that innovation from the business sector into this ICT sector because I still feel somehow something is holding us back. Where we see some of the unregulated business sectors highly innovative with all sorts of new business models, Google coming up with a phone, Apple coming up with the iPhone. With the Grameen Bank microlending moving to the Grameen phone. The unregulated business sector is highly innovative. I think I still get some of that feeling that we are still being held back by something. And I can't quite put the finger on it. My feeling and my fear is that perhaps that sector in itself as it is currently still highly regulated, high barriers to entry through licencing systems, high investment cost. Is it an economic problem in the sector? Is it a regulatory problem in the sector? How can we move that sector from kind of those existing incumbent operators or those who have found that is a nice place and try to protect their existing market and their income and their established business model? How can we move that into a more dynamic innovative sort of a fashion so that we can connect all those fantastic things that Prof. Marsden brought up, these innovative things that come up from the bottom? How that can be transformative similar to Prof. Yunus' Grameen model has become mainstream now that these kinds of things become. Maybe Mr. Heuermann, you know about the regulatory environment in all of those countries. Maybe you can start and maybe Sania? Thanks.

Prof. Picot:
This is a major topic that you are addressing. I think, Dr. Heuermann is the right person to give a first answer and then also Mr. Scherf who is consultant in some of the developing countries improving their regulatory environment could also comment on this. Of course, everybody else is invited. Please, go ahead Dr. Heuermann!

Dr. Heuermann:
Thank you. First of all Mr. Hermans, I think CSR (corporate social responsibility) has come into mainstream of business, already. For example if my company is writing a Bid Book for an investor who wants to buy in a privatization deal we always introduce a chapter on cooperate social responsibility. Many CEOs meanwhile have seen the advantage of being a good citizen and having a good

image in the country not only for delivering good service and also in the social sense of the word. How can we get innovation into the sector is a very difficult, question. First of all I would say the technical development to a next generation network will lead to two ways in how these sectors are developing. On the one hand side we will have a more standardized network, the all IP "next generation network" which will be very similar everywhere in the world. And managers of such a network are those have to follow a cost leadership strategy and offer a commodity to others. On the other hand the NGN will have the opportunity of a separate application and service layer, where providers can offer very individual services. Here we come to the innovation side and the more cultural adaptive part of innovation. I think this can already be seen in European countries where we have already reached saturation in many cases and dozens of MVNOs have emerged for example in the mobile business. These are not network operators, but service providers which are offering customized services to segments like young people, like the Turkish minority or like cheap "no frills" offers. For each of the customer segments you get very specific solutions in different languages for different cultures, for a specific live span, for elderly or for young people. I think we will see the same development very soon in the developing countries. The only difference here is they are still in the high growth phase. And if you sell water in the desert you don't differentiate your product. This is the same in the motor industry. 40 years ago we had all the people driving in a VW "beatle" in the country and that was a mass product. Today, I think Volkswagen alone has ten different brands and each brand has I don't know how many different cars; a very segmented market. We will see the same in the ICT industry in most projects. Service development in a NGN network is very cheap and easy. That is another argument for having the possibility at least also in each of the developing countries to innovate in services, not so much on the network side but in services.

Prof. Picot:
If I may intervene here this would also demand that the service layer is dis3ntangled from the infrastructure layers so that really the service can flourish without having to take into account certain bonds with the infrastructure. One of you mentioned that we have seen more of vertical integration which just runs counter to this development and perhaps you could take up this aspect and other aspect within the regulatory environment.

Mr. Scherf:
One comment on the CSR issue which Mr. Hellmond raised. What we have learnt out of the telecentre movement in the 90ies is that financial rentability is the key for a sustainable provision of ICT services and infrastructure in developing countries. There definitely has to be a business case in order to reach sustainability. An example: Some Years ago I visited a multi-purpose community center in Uganda after after ITU's subsidizing ended. The only application which was still running was the copy machine and nothing else. In the case of regulatory issues I think there

are two aspects. First, it is important to have an adequate regulatory and legal framework. Second, these frameworks have to be operational and effective. And this is the aspect where many regulators in developing countries are struggling. Due to capacity constraints they struggle with things like process management or change management. Even if there are adequate regulatory and legal frameworks but due to lack of personal or financial capacity regulators face enormous problems to get them operational.

Ms. Melhem:
If I can add to what you just said. I completely agree with you and regulators are part of governments in many of the countries. And still I go back to my poit. There is a big contrast because how backwards a lot of governments and public sectors is because the bright people are going to be employed in the private sector or are simply emigrating in a lot of countries we see. And I want to add one thing to the point because there is a lot of political economy that is still out there. A lot of these operators are still the babies of the state. They are protected by the political structure and hence this changes the context for competition. These state owned operators are big employers. I can tell you I work on Middle East and Africa. We call a lot of these operators failed incumbents. They employ around 25.000 people. They are so inefficient. Just looking at the number of subscribers and the number of employees they should be slashed by the hundred. Yet, it is employment. It is political. Some of these countries with fixed line incumbents have been losing clients by the thousands, even by the millions. Their companies are left neglected, their assets like the copper line infrastructure are left neglected, they would be better off restructured, and operated by the private sector while the employees who are not retained and trained on other skills that are really adequate for employment in the context of the country, like hardware and network maintenance amongst others or customer support for DSL or other technologies

Prof. Picot:
What is your addition to this?

Ms. Esterhuysen:
I think it does require a mind set shift as you are saying. I agree with that. I think in particular at two levels research and development, and how that is thought about, and also partnerships with the public sector. Many of the large foreign ICT companies initially went into a public private partnership as a way of bypassing a very restrictive regulatory environment. It was often a way of getting into that market. It was often not a very healthy partnership. I think really re-thinking those partnerships. The regulators often won't have the capacity to regulate effectively. But looking at making them sustainable, you see this with some companies and some of the initatives we heard about today. If you are going to invest in a healthcare initiative you partner with the public health system, you make sure you got human capacity, evaluation and monitoring. There is a qualitative way in which

cooperate social investment can make a big difference, even if it is for profit. And when you think at research and development, the output or the goal of R&D is not only to look at tapping new markets and creating new consumers. But also looking at creating producers. This is the line we have been creating more sustainable global economy. With a globalized economy you are so much more interdependent on one another economic health. So, I think it makes sense. I really like the example we heard from the Microsoft research lab where they did things by the projects. And then we have theses three different types of output. I am not quite sure if I remember them correctly. But in some cases the initiative was up scalable. There was a business case for it. And in another case a research article was produced that can help others learn from those experiences. In another case a not for profit initiative was being established in partnership with a NGOs and local government. If that kind of thinking can really falter through R&D, then what cooperations invest in R&D could in fact have longer term social benefits, as if it is not narrowly focused just on increasing profitability in a small way.

Prof. Picot:
Thank you. Dr. Wenzek, please.

Dr. Wenzek:
When I am listening to the original question of how the rollout can be handled and the responses, I would like to make a recommendation and hear feedback based on your respective application domains. I recommend that we should try to solve a complete problem.

Surely, we need to continue to build out the infrastructure: roads, ICT, utilities etc. But all of that is just still the enabling function for something that you really want to do, that is relief people from poverty, educate them, provide them with healthcare. We need to find e.g. three different clearly stated scenarios that we can tackle together in a public private partnership, starting with one country first. At IBM we have a strategy communications principle around the *Smarter Planet*. As part of that we are looking into a project called "How do you make the planet smarter? One country at a time". This way you can start to replicate based on real learning. But to get that learning one needs to address a complete problem, not just parts of it. Let me give you an example where a problem is addressed in its completeness, one that you have probably heard of already: Electric vehicles and Shai Agassi's *Better Place* organization that tackles the challenge of transforming individual transportation from gas to electricity that is produced from renewable energy. They always address a small countries like Israel or country-like states, California, Hawaii, etc. to solve the whole problem from provisioning electricity, providing the vehicles, the batteries, the loading stations etc. They build a solution for the whole system, but in clearly defined, smaller boundaries.

So far we were talking mainly about devices – or cars. But you can find any application domain and then go out and describe the whole solution scenario about it. Then bring the right parties together and address this complete problem for a clearly defined country or region. Probably by just focusing on the people in this small room we could define 1-3 different scenarios. We should get together and set the projects up by defining what it takes us to solve that problem for the whole population of one specific country. Then do the research in such a vigorous way as explained by the Microsoft colleague so it can be published and accepted by an academic, independent community. This way it is not only the business acting, but academia would be able to add credibility and find other ways to replicate the solution in many ways.

So let's discuss what the three top applications domains to tackle might be, figure out what the right countries are and then get it done.

Prof, Picot:
Is this a good advice for the future strategy of the Worldbank?

Ms. Melhem:
I think this is an excellent advice and it comes from a problem solver not a politician. We tried to have this approach on a few post countries, Cambodia is one that comes to my mind. After a big war everybody is here, everybody is opening their pocket. The new government is open. So, it always starts like that in my experience. And then the doner wants to fund agriculture, this doner wants to fund foreign trade and thinks fragment. Then within the same donor community, we have a group that helps the health sector, another group that helps social security and we found the same data basis for the same citizens with different systems that don't talk to one another. We don't plan financings in maintenance and then the project finishes after three years. It is that short term fragmented approach. I am sure my donor colleagues here on the table would say the same. Or from the private sector very fragmented. I agree with you. It is about time I think after all these years we are still looking for success stories to say okay, this is not the right way. How can we do it differently? This is a whole political aspect but I completely agree with you. We should have a holistic country based approach and then we could avoid some of the problems that we have, nor the costs that we have. But where to start when all the countries need the help?

Prof. Picot:
We have arrived at the end of our schedule this afternoon. Let me at this point first of all thank all my colleagues on the panel for their very interesting and stimulating inputs and also the open discussion. I think this final round has shown again that we have tackled a valuable subject on this conference and at the same time that there are quite a few very difficult questions to be delt with in future. Especially let me just pick two or three of them.

The wholesale approach as a very major strategy for network providers in the various countries, not only in developing countries but also in other countries. It should be really much more unfolded in order to bring services to all the corners of the regions and at the same time make services blooming. But this also is very tightly intertwined with the fixed mobile convergence trend that we are seeing and that will continue to take place. So, we have to acknowledge a saying that I learnt a couple of weeks ago and which goes: the future of mobile is fixed. So, the mobile systems in this world cannot really grow in future and at the same time take up the demand for capacity if they are not linked to a very powerful backbone network which comes close to the points of demand and points of consumption. They need a very powerful fixed network in order to rollout their services in a flexible mobile fashion that is of course demanded by many consumers and businesses. Thus, we need regulatory conditions and incentives to really push forward the fixed line implementation and investment. Otherwise we will see a situation where the digital divide will even increase, and this is not what we want. It is a very difficult task but a very urgent one not only in the developing countries but also in other countries. We learnt that some European countries see a similar structure where the mobile subscribers by far exceed the fixed subscribers. This is not sustainable. At least if we want to provide high quality and multimedia solutions and services in future.

We have also learnt that it is possible and needed, perhaps needed more badly, to come up with cultural specific and adapted solutions and services and that perhaps the key for success in those markets lies in the close collaboration and communication with the people on the spot. This is also one of your messages, Dr. Islam: "Get down and dirty" as you called it. I think this slogan is well taken. It shows that we have to be very close to those who might have demands and surely have demands but we have to understand them and to transform them into sustainable solutions that then can be provided by infrastructure based services that are being rolled out.

These are only some of the many aspects that we have been able to touch during the last 60 minutes. There was more, but I might also remind you that we will come up with a documentation of this conference. So, if you want to re-read the arguments you will be able to do so. I might also mention that the slides that were shown today will be available on the web-site of the Münchner Kreis within in the next few days so you can access them and even download if you want to do so. We are all caused to digest the many aspect that we have swallowed in today and then perhaps come up with a very good final product.

12 Closing Remarks

Prof. Dr. Arnold Picot,
Ludwig-Maximilians-Universität, Munich

Ladies and Gentlemen, let me at the end of this conference thank those who have made this conference possible, especially Josef Lorenz and Jörg Eberspächer and a program committee that has prepared this conference. I think they have done a great job. Thank you very much.

At the same time I would like to say a warm thank you to all the speakers and discussants and to all of you who have come to this conference to Berlin today and discussed with us. I think we were not a very large but a very competent audience. I feel that it was really worth coming here and that we got food for thought which will hopefully be transformed into an action be it in the political scene, in the business scene or in the scientific scene.

This is one of the missions of the Münchner Kreis and I hope that we will live up to that. Thank you very much and I wish everybody a good and safe way home.

A. Picot and J. Lorenz (eds.), *ICT for the Next Five Billion People: Information and Communication for Sustainable Development*, DOI 10.1007/978-3-642-12225-5_12,
© Springer-Verlag Berlin Heidelberg 2010

Appendix

List of Speakers and Chairmen

Prof. Dr. Joachim von Braun

Director General
International Food Policy Research
Institute
2033 K Street, NW
Washington, DC 20006-1002
USA
j.vonbraun@cgiar.org

Jean-Marc Cannet

Manager Strategic Customer
Engagements
Alcatel-Lucent
7-9 avenue Morane Saulnier
78141 Velizy Cedex
FRANCE
jean-marc.cannet@alcatel-lucent.com

Dr. Stanley Chia

Senior Director
Vodafone Group R&D
2999 Oak Road
Walnut Creek, CA 94597
USA
Stanley.Chia@vodafone.com

Prof. Dr.-Ing. Jörg Eberspächer

Technische Universität München
Lehrstuhl für Kommunikationsnetze
Arcisstr. 21
80290 München
joerg.eberspaecher@tum.de

Anriette Esterhuysen

Executive Director
Association for Progressive
Communications
PO Box 29755
2109 Melville
SOUTH AFRICA
anriette@apc.org

Dr. Arnulf Heuermann

Detecon International GmbH
Oberkasseler Str. 3
53227 Bonn
Arnulf.Heuermann@detecon.com

Kazi Islam

CEO
Grameen Solutions Limited
Grameen Bank Tower, 12th Floor
Mirpur 2
Dhaka 1216
BANGLADESH
Kazi.Islam@grameensolutions.com

Josef Lorenz

COO RTP Innovations
Nokia Siemens Networks GmbH
St.-Martin-Str. 53
80240 München
josef.lorenz@nsn.com

Prof. Gary Marsden

ICT4D Centre
University of Cape Town
Cape Town
SOUTH AFRICA
gaz@cs.uct.ac.za

Samia Melhem

Senior Operations Officer
c/o The Worldbank Group
2121 Pennsylvania avenue NW
Suite F5P-136
Washington, DC 20433
USA
smelhem@worldbank.org

Christian Merz

SAP AG
CEC Karlsruhe
Vincenz-Priessnitz-Str. 1
76131 Karlsruhe
christian.merz@sap.com

Frank Oehler

Head of Business Development
New Growth Markets
Nokia Siemens Networks
Karaportti 3
FI-02610 Espoo
FINLAND
frank.oehler@nsn.com

Prof. Dr. Dres. h.c. Arnold Picot

Universität München
Institut für Information, Organisation
und Management
Ludwigstr. 28
80539 München
picot@lmu.de

Thorsten Scherf

Bundesministerium für wirtschaftliche
Zusammenarbeit und Entwicklung
(BMZ)
Referat 300
Adenauerallee 139-141
53113 Bonn
thorsten.scherf@bmz.bund.de

Dr. Kentaro Toyama

Managing Director
Microsoft Research India
"Scientia"
196/36 2nd Main
Sadashivnagar, Bangalore 560 080
INDIA
Kentaro.Toyama@microsoft.com

Dr. Hagen Wenzek

Corporate Strategy
IBM
1 New Orchard Road
Armonk, NY 10504
USA
hwenzek@us.ibm.com

Lightning Source UK Ltd.
Milton Keynes UK
19 July 2010
157210UK00003B/130/P

05165970